Scientific and Prophetic Evidence for the Christian Faith

A Concise Handbook of Evidence for Christian Belief and Witness

James L. Lowther, PhD
(The author of the Danny Carter Series)
2014

0

CONTENTS

1

INTRODUCTION

Every fall millions of teens head off to college with high hopes and dreams. Some want to be teachers and some engineers or businessmen. All of them want success, and perhaps, to have a little fun along the way. What many unsuspecting Christian parents do not realize, however, is that many colleges and professor have an unannounced goal to knock any faith that a student has out of him or her.

Being away from home for the first time is a wondrous thing, but it is also dangerous. There are many pitfalls out there, from drunken parties and drugs to promiscuity and other students looking to take advantage of the unsuspecting novice.

The saying is that to be forewarned is to be forearmed. This little book is designed to help arm the Christian student and parents against those who would try to rob them of their faith. There are so many good answers to the critics of the Bible and Christianity that it is a shame that our youth are primarily not equipped with those answers before they go off to college or even in public high schools.

One naiveté that plagues our youth is that there is such an aura surrounding the position of professor, that students swallow everything that he says without any questioning. After all, he is the expert in his field. There is no concept that he is a fallible human being and may have a less than pure agenda to promote. We need to tell our young people to listen with discernment and use the Bible as a grid in which to evaluate what is being taught.

This work is obviously not designed to provide all the answers, but it is designed to help students and others think logically through the great issues of faith that will be challenged before the challenge is given. Not only are there

3

good reasoned answers available to defend against attacks on the Christian faith, but those answers are not on a seminary level. In fact, naturalism and evolution fail on the elementary science level, as you will see. Also, the Bible is easily attested to by many of fulfilled prophecies. You can, indeed, be confident in your Christian faith and can successfully defend it from attack.

I also hope to stimulate your thinking and hope that you would go ahead and do more reading and research on your own. I would suggest reading *Darwin's Black Box* and *The Edge of Evolution* by Dr. Michael Behe, *The Privileged Planet* by Gonzales and Richardson, works from The Creation Science Institute, and, of course, the first two books in my trilogy, *The Group* and *The Debate.* I would also encourage you to read *Evidence that Demands a Verdict* and *More Evidence that Demands a Verdict* by Josh McDowell. At the very least, view the DVDs *The Privileged Planet* and *Unlocking the Mystery of Life.* Finally, go to YouTube and view the short video on Fibbonacci's number. Prepare to be amazed. Hopefully, you will find this little succinct work both helpful and enjoyable. Keep in mind that the works by Behe and Gonzales are not from an evangelical perspective. However, these scientists have honestly considered the issues at hand and have concluded that an intelligent designer had to be at work in order to produce the universe and life.

Finally, let me say a word about the difference between this little manual and my books, *The Group* and *The Debate.* In this manual you will receive a summary of the evidence supporting the Christian faith. In the trilogy, *The Danny Carter Series*, you will experience in real life depiction a Christian defending his faith on a secular campus. These novels are designed to prepare the Christian student for what criticisms he or she will encounter, and how to deal with them. Also, the trilogy will give a flavor for the price Christians might pay for standing for their faith. One more difference between this preparatory manual and the larger works, is that the trilogy has more detailed information in them.

I hope that you will agree that the information provided in this work is both informative and entertaining. May the Lord bless your walk with and service to Him.

Chapter 1
Science 101: It is really not that hard!

I enjoyed going out to lunch to Mama Stella's, but not when the cost was so high. The builder of our new church always took me to this restaurant when he was going to hand me another $100,000 bill for work he had already completed on the building. I hardly thought that lunch for a hundred grand bill was quite a fair trade, but since he had done the work on the building, he deserved to be paid.

After a while on one occasion the conversation turned to evolution. The builder had inquired about what kind of church we were, and I told him that we were an independent Bible teaching church.

"Well, if you believe in the Bible, then what do you do with evolution?"

"There is far more evidence against evolution than for it. For example, all the parts of a cell and all the vital parts of an organism must exist at the same time in order for there to be any life at all. These complex inter-working systems could not come about by gradual random processes."

"Well, I don't know about such things. I just know that the experts tell us that evolution happened."

"I could show you the evidence that tells us that creation is the only answer to existence."

With a wave of the hand the builder indicated that he didn't want a discussion on the matter, even though he was the one who brought the issue up. "You have studied such things. I have not."

Not only had the quite intelligent engineer before me not studied such things, he had no interest in pursuing the matter. The subject had profound implications for his soul, but it held no interest for him. Unfortunately, his reaction was all too typical of that of the vast majority of people.

6

As a society, we are so used to listening to the experts, that we rarely question them, even when what they are saying makes no logical sense. For example, if you watch debates between naturalists (those who believe in accidental random origins of all things and evolutionary theory) on YouTube, you hear some fairly outrageous pronouncements that assault the intelligence of the listener. However, since we are accustomed to deferring to experts in the field on such matters, we tend to take what is said at face value without further thought or investigation.

We tend to make the assumption that science is just too deep for us to understand. We must allow the trained smart people with lots of initials after their names to dictate to us what reality looks like. Added to the mystique behind these pronouncements is the full backing of the media and other experts in other fields who accept the naturalistic experts as true scientists with an abundance of evidence to back up their claims. Even when we know that a mock up image of an extinct animal is produced from one fossilized tooth, we still assume that there must be solid evidence behind the drawing or sculpture.

Allow me to share with you some of the actual comments made by leading naturalists. Dr. Peter Atkins has stated that in the beginning was nothing, but nothing was made up of positive and negative charges which split apart to make the universe.

Huh? Talk about a head scratcher! Yet, here we go again. Many people will assume that since a very smart PhD in chemistry made the statement, it must be true. But, reader, don't sell yourself short. If you know that water is wet, don't let someone talk you out of it. If a concept breaks down at Science 101, it cannot be revived at Advanced Physical Chemistry 411.

We know several things that elementary science teaches us. First of all, nothing is nothing. That is, it is the absence of everything. Aristotle said that nothing is what rocks dream of. So, if nothing is the absence of

7

everything, then it cannot be made up of anything. To put it another way, if I add 0 to 0, I come to the sum total of 0. 0 is made up of nothing.

The second thing we know, dear reader, is that positive and negative charges are attracted to each other. So, if the universe split apart, it would not have been through the dividing of positive and negative, since positive and negative cling to each other like gum on a shoe. Anyone who has played with magnets knows this concept (it is only by the ingenuous design by God of strong nuclear force that prevents electrons around atoms from collapsing into its positively charged center. This force is so precisely set, that any variation in strength would make our universe uninhabitable).

The third thing that we know is that stating that the universe started out as positive and negative charges begs the question as to where these charges came from to begin with. All naturalistic systems suffer from this same fatal error. They all start out with matter, but do not explain how the matter came into existence.

Dr. Richard Dawkins's statement that earth could have been seeded by aliens, but those aliens would have evolved illustrates the point very well (see the DVD *Expelled: No Intelligence Allowed*). To deal with origins requires an explanation of the beginning of all things, not just the beginning of things we know.

I will use one more example. Dr. Stephen Hawkings, often recognized as the most brilliant mind of the last part of the twentieth century and beginning of the twenty-first century, has stated that in the beginning was gravity. Gravity then created all things. Here is another *huh?* moment. You might say, "Well, I thought gravity was produced by massive bodies, like the earth." Don't throw away that idea just because a smart man has said something different from what you were taught. All gravity that we know of is directly related to the mass of the body exerting the gravity. The more the mass, the greater the gravity.

8

The smaller the mass, the less the gravity. Therefore, if in the beginning there were no mass, then there was no gravity either.

Wow, if you think that the pronouncements by these scientists are mind boggling, wait until we come to molecules riding on the back of crystals later in the book. Oh, yeah, I almost forgot. The multiple universe (that there are an infinite number of universes out there, so there are an infinite number of possibilities, so all things are probable--what utter nonsense!) theory is also a thriller.

Chapter 2
Is naturalism and evolution science or philosophy?

I watched the CNN interview with fascination. Here was a physicist, Dr. Lawrence Krauss, and the microbiologist Dr. Michael Behe being interviewed. The question of the day was whether Intelligent Design should be allowed in the science textbooks to be taught in the public schools. Krauss argued that the ID proponents had to spend decades writing scientific papers that would then be scrutinized by the scientific community before such a thing could be allowed a place in the textbooks used by the nation's children.

Dr. Behe asked what Krauss and others like him were afraid of. The ID proponents had certainly demonstrated the viability of their positions in articles and in such books as *Darwin's Black Box* and *On the Edge of Evolution.*

I thought as I watched the interview how the prevailing scientific community was desperately trying to hold off any view that was not their own, for the majority view just did not hold up well under close examination. The fall back position was to claim that ID was merely an attempt to introduce creationism into the classroom. Ironically, most of the ID proponents do not even claim to be Bible believing Christians.

The Intelligent Design proponents have caused quite a stir over the last few decades. As long as evolutionists could dismiss creationism as just the archaic wishful view of a bunch of backward Christian fundamentalists, they could claim the scientific field all to themselves. Even though many qualified creation scientists were on the scene, they were simply ignored or mocked.

Then out of the blue, a group of respected highly credentialed scientists (Behe, Kenyon, Gonzales, Richardson, and others) announced the new concept of Intelligent Design as a reasonable explanation of origins.

These scientists had given up on naturalism as an explanation of existence due to three factors. One, life is irreducibly complex. That is to say, all the complex inter-working of cells with incredibly fine tuned micro machinery had to have an awesomely powerful, intelligent, and wise engineer behind them. Two, life is interactively balanced in an irreversibly intertwined biosphere (life depends on other life and finely tuned environmental factors to live). Third, the earth itself is highly fine tuned for life, with such a small margin of error in its location and make up, that the earth of necessity had to be engineered specifically to support life.

This brings up the question, then, *is the study of origins science or philosophy?* Bill Nye (*The Science Guy*) has stated that Christians should not teach creationism to their children because the nation needs scientists. He speaks as if naturalism were not merely the foundation of naturalistic philosophy, but of science itself. It is absurd to think that one could not be a biologists, physicist, astronomer, or chemist without believing in evolution and naturalism first.

To answer the stated question (is evolution science or philosophy?) we must first ask, *what is science?* The word science comes from the Greek word for knowledge. Therefore, science investigates certain knowledge about our universe, particularly the earth and our solar system. Science is broken down into disciplines (chemistry, biology, physics, geology, genetics, etc.). The factual knowledge of science is gained through experiments, measurements, and observation that is repeatable by other scientists. Anything that is not observable, measurable, or repeatable is not science. For example, we know the freezing and boiling points of water at various altitudes, for we can demonstrate these facts over and over again. This is science. However, we have never seen the origins of the universe or life, so the study of origins is not science. It is philosophy. Theories and hypotheses are just that. They are

11

educated guesses, though some are merely pure speculation.

Does this mean that nothing about the study of origins is science? Not at all. Actual scientists who truly lay aside all preconceived notions can observe, experiment, and measure what actually exists, and then eliminate any theories that could not arrive at what they see. Especially since the invention of the electron microscope in the middle of the twentieth century which allowed for a detailed look at the cell, scientists are able to discern what could or could not have happened.

In spite of this enhanced ability, no one can say with certainty from a purely scientific point of view what actually happened at the beginning of all things. Since this is true, all speculation of how matter and energy began falls into the category of philosophy. This means that the discussion of origins has no place in biology, physics, chemistry, and other science books. It belongs exclusively to the field of philosophy and should be kept in this discipline.

Ah, but you may protest that evolutionists and developmental biologists are scientists. In this regard you would be right, but when scientists ascertain anything about origins they are speaking as philosophers and not scientists. This is particularly true when what they proclaim does not match up with the laws of science.

Then, should evolution be taught in biology books? The answer is an emphatic no, unless one would like to include a philosophical section, labeled as such, that includes all major theories in the discussion. To treat philosophy as science is dishonest and does a great disservice to our students and the public at large. For this reason, the Apostle Paul warned Timothy about science falsely so called (1 Timothy 6:20).

12

Chapter 3
It's a big universe: how did it all begin?

The famous astronomer, Carl Sagan, stood before a huge star map produced from a picture from the Hubble Space Telescope. He looked into the camera lens and profoundly proclaimed, "This is all there is and all there ever was." This was quite a statement from someone who could not possibly know all that is, let alone all that ever was.

I would not call such a bold assertion arrogant, since that would be rude. Let's just say, the statement is misinformed or misguided. No one knows all that ever was, except God. What He has revealed to us is but a snippet of the entire breadth of reality, therefore I would not dare to make any proclamation of all that was before we came on the scene.

Instead of delving into what was, let's look at what could or could not have been. By eliminating what could not have been, we can obtain a better picture of what must have been. Allow me to illustrate the point. Let's say I run across a hubcap on the ground in the woods with the word Ford emblazoned on it. Since the hubcap obviously has form and function, I can eliminate the possibility of it being an accidental formation of nature. Also, I can eliminate that it grew there from some hubcap plant, since such things do not occur in nature. I must assume, then, that an intelligent designer made the hubcap for a particular purpose and somehow this designer was connected to the name or label Ford.

Now, let's look at the same logic when we come to the universe. The prevailing theory of the universe is that it is very old (on the order of 10-15 billion years old) and was initially formed from some explosion of a lot of hot gas particles. After the explosion (called the event horizon by scientists),

13

matter spread out all over the place, spinning and coalescing into galaxies, stars, and planets, until what we see today came into being. Ironically, the atheistic philosopher, Chris Hitchens, calls this the miracle of the universe. Of course, it is hard to imagine a miracle without a miracle worker, but let's move on.

As we look at animated computer images of this process, we are amazed. Wow, what a wonderful thing. We see gasses and explosions, and spinning material, forming into galaxies and stars and planets--quite amazing stuff. Here is the problem: is it true?

To see if something is true or not, we must first see if it is false (this is the test of falsification). For the universe to have been created by such an imagined process, the process itself must obey the laws of physics as we know it. Let's put this model to the test then. Don't worry; we will not get too deep. In fact, we will not go beyond elementary science and good ole common sense.

First of all, where did all this gas come from to begin with? At this point we hear crickets in the darkness, since there is no answer to this question. In chapter one we discussed some proposed explanations which so defy the laws of physics, that they are not worth considering.

Secondly, everyone knows that explosions create messes and not galaxies. As we look on in horror at the mess left behind by terrorist attacks, we realize that all explosions tear things apart and not put things together. Why do you think they call men who bring down buildings demolition experts? They demolish things (which for a guy is more fun than building things, but that's another issue for another day).

Third, an explosion sends material out from the center of the event horizon (starting point) uniformly in all directions. This would produce a spherical pattern. This does not describe our universe, for our universe is very,

14

very flat. In fact, it is so flat, that if looked at on edge it would look like a thin paper plate, for while the universe is 20 billion light years across (light travels at approximately 186,000 miles per second & a light year is the distance light travels in one year), but only 200,000 light years thick. The entire universe, then, is on a massive disc, just like our solar system is.

Fourth, the universe is actually quite young. It has to be, since the energy dissipated in the universe would have run out long ago if it were very old. But don't take my word for it, ask the moon. No, I did not say ask for the moon--ask the moon. When the astronauts Buzz Aldrin and Neal Armstrong landed on the moon in July, 1969, their lunar module was equipped with large pads so that the module would not sink into the hundreds of feet of fine dust that would have collected on the surface over billions of years of existence. They needed not to have worried, for they found less than six inches of fine dust on the surface--a half a foot. In fact, the fine dust was so shallow that the astronauts had a hard time making the American flag stand up straight, for the pole just did not have much sand to go into. So how long would it take six inches of dust to accumulate on the atmosphere deprived moon? -- about 10,000 years or less.

Now just a nanosecond here. You might protest that since we can see distant galaxies that are billions of light years away, and since it takes billions of years for this light to reach us, the universe must be billions of years old. This looks to be apparently true. However, since Genesis 1 tells us that God created light on the first day and the sun and stars on day 4, we know that God created the universe with an appearance of age. That is to say, since God wanted us to see His glory in the heavens (Psalm 19), He made it so that we could see the vast objects in the universe immediately instead of waiting billions of years to see them. Since there was no way that Moses, the author of Genesis, would have known this on his own, we have good confirmation that

15

God meant for us to see these great wonders instantly instead of waiting for light to make its way to us from the outer reaches of our universe.

The fifth problem deals with gravity. In elementary school we learned that dropped objects fall toward the center of the earth. Of course, we did not need a teacher to tell us that. We already knew by experience about spilled milk, dropped eggs, and dropping heavy books on our toes. We then learned that the more massive an object (dense), the greater its gravitational pull. Finally, we learned that the closer two objects are to each other the greater their attraction to each other, and the farther objects are away from each other the lesser their attraction to each other. In fact, gravitational attraction drops fairly rapidly as objects move away from each other.

Taking this elementary understanding, let's apply this knowledge to the proposed Big Bang theory. As this huge ball of gas compresses it reaches incredible temperatures of billions of degrees. When it can compress no more, it explodes in a tremendous crescendo of energy. Within seconds particles are flying in all directions heading away from the center and from **each other**. After only a few seconds each particle is so far apart from each other, that there is no measurable gravitational attraction between them at all. In fact, the law of inertia (objects that are at rest tend to remain at rest unless acted upon by an external force, and objects that are in motion tend to remain in motion along the path traveling unless acted upon by an external force) does not allow any of the dispersed particles to do anything but to continue to travel along a straight since there are no other forces to act upon it.

What does this mean? It means that there can be no coalescing into galaxies, stars, solar systems, or anything else for that matter. A Big Bang would not only produce a big mess, but it would also produce a never ending spherical dispersal of particles so far apart as to be nothing but individual small pieces of matter in a vast void.

16

Sixth, Venus is a problem. Why is Venus a problem? It's backwards, that's why. When anything explodes, the particles are sent spinning in the same rotation to each other (called angular momentum). In other words, if the force were such as to spin a particle clockwise (right to left for digital watch fans), then all the particles would spin in that direction. If the particles spin end over end in a forward direction, then they all do. Since Venus spins in the opposite direction of all the other planets in the solar system, Venus and earth did not come from the same explosive event. Oh, by the way, just for the fun of it, Uranus spins on its side in relationship to its orbit.

Seventh, there is a baldness problem. No, I'm not talking about going to a hair specialist. Earth does not have a baldness problem, so there is a problem for naturalists that the earth is not bald. You see, if the earth were as old as naturalists claim that it is, then due to soil creep (erosion), the earth would have worn down and become flat a long time ago. Gravity just keeps pulling on the surface of the earth (and us -- we actually compress and get shorter as time goes on). The fact that earth is still hilly, means that it is still quite young (this can be seen in the collapse of the Boy Scout Cave in Craters of the Moon, Idaho, and the continuing collapse of the rock arches in Yosemite, since if the earth were old, there would be no caves or rock arches [all of them would have collapsed]).

Finally, material itself is a problem. Since earth is made up of different material than the moon, Jupiter, Venus, and the other planets (not to mention the sun), earth did not come from the same source that the other planets came from. This means that a single cosmic explosion could not have produced what we see today.

As we have demonstrated, the Big Bang does not pass the test of falsification on any level, therefore it is not a candidate for being the originator of all things (I'm going to go with Genesis 1 on this one). The reader, therefore,

17

can safely discard this theory of origins. Also, since wrong premises lead to wrong conclusions, everything built upon the wrong assumption of the Big Bang is also suspect. Finally, if naturalistic theory cannot pass the muster of elementary science, it can climb no higher. The theory must be jettisoned for a better one.

18

Chapter 4
Red shifted and red faced

Dr. William Benson was a brilliant scientist and his loving wife, Dorothy, loved to listen to his explanations of the progress of his work, even though she understood little of it. Dorothy was an elementary school teacher with a good mind, but was far more adept at basic concepts than she was at complex details. When Bill came home on one particular Friday evening, she knew that something was bothering him.

"Bill, what's wrong?"

Bill was obviously preoccupied with whatever was on his mind. "Huh? Nothing. Why?"

"Come on, Bill. I have not been married to you for twenty years without knowing that something is troubling you. Now out with it."

"Okay. Everything is red shifted."

"Uh, what?"

"Everything is red shifted."

"What do you mean?"

"You know that train track at the bottom of the hill?" Dorothy nodded. "When the train reaches the crossing at Wilson Road the engineer blows the whistle. As the train approaches the crossing the whistle is high pitched, but when the train passes the pitch drops."

"I know. It's called the Doppler effect. The sound waves are compressed as the trains come toward you and then stretched out as the train moves away from you, due to the movement of the train."

"Right. Well, light does the same thing. As galaxies move toward us, having been formed after our galaxy formed, the light is blue shifted on the spectral scale, since the light is compressed. If a galaxy is moving away from

19

us, then the light is red shifted due to the light being stretched out."

"I see. And you say that everything is red shifted from us?"

"Yes."

"What does that mean?"

"It means that all the galaxies are moving away from us and none are moving toward us."

"Doesn't that make us near the center of the universe."

"That is what it would imply, but that just can't be. The center has to be empty and some galaxies must be behind us and some ahead of us."

Dorothy shook her head. "Then what is your explanation for this phenomenon?"

"I don't have one yet."

"Does this indicate that the Big Bang theory is wrong?"

"It can't be. It just can't be."

As we can see by this look into the life of an astronomer, evidence keeps getting in the way of preconceived ideas. The idea that there was a big explosion of matter at the beginning of the universe that formed galaxies and everything else would not allow for us to be at the center of the universe. Since everything that we can see through our telescopes are red shifted from us, then we are indeed at the center and, therefore, the Big Bang could not be an explanation for the origins of the universe.

Often naturalists and evolutionists are stunned by discovering things that do not fit into their concept of the way things must be. After all, for their theories to hold water, the evidence must somehow fit. However, time after time, these scientists run across things that just do not fit their preconceived notions. When this happens there are only five options: 1. admit that they were wrong, 2. explain away the new findings as an anomaly, 3. invent an explanation that sounds plausible, even if it is either unlikely or impossible, 4.

set the findings aside to be dealt with another day, or 5. ignore the findings as if they were never discovered.

As to the first point, obviously most scientists are not going to give up their belief systems in favor of another system contrary to their beliefs. So, this option just is not an option at all. Points two through four are far more popular among naturalists. Through computer animation or a dismissive wave of the hand, the scientist puts a patchwork on the hole created in his theory and then moves on. If this does not work, he simply says, "We are working on that." However, even those who do such things tend to use point five in their own literature. That is to say, he just ignores the issue all together.

There is a concept in all fields of study called a priori assumption or axiomatic foundation. All this means is that there are certain starting points that are assumed to be true without need of proof. In math we take for granted that the distant between one and two is exactly the same distance as between eight and nine. In English we do not set out to prove that a noun is a person, place, thing, or idea. We merely accept this concept as fact.

However, if our a priori axioms (our starting points) are wrong, then everything built upon them will be wrong also. It was once universally assumed that the earth was flat and that the universe revolved around it. When this was shown to be incorrect, those who firmly believed these concepts threatened to slay those who had demonstrated that the foundational assumptions were wrong.

Now, if there were no Big Bang (a derisive term coined by astronomer Fred Hoyle), then the entire foundation of modern developmental astrophysics comes toppling down. Once the foundation crumbles, the entire theory must follow, even if thousands still proclaim it to be otherwise. No matter who proclaims how beautiful the emperor's garments are, if he is actually wearing no garment, the emperor still has no clothes.

21

Chapter 5
Earth: nice neighborhood (it's all about location, location, location)

The church that I pastor is just outside Joint Base Andrews (formerly Andrews Air Force Base) in Maryland. When the 9-11 terrorist attack occurred in New York and at the Pentagon, the people in my church who lived on base wanted to get back on base for safety. They figured that with guards at the gate that would be the safest place to be. However, the civilians in my church wanted to get as far away from the base as possible, figuring that the base would be a prime target for an aerial attack, particularly since it was the presidential base.

Both sets of people were looking at the location of the base in a different manner. The same is true when it comes to scientists looking at the position of the earth in the universe. The medieval church, using Ptolemy's model, taught that the earth was the center of the universe and everything revolved around it. Then along came Galileo, Kepler, and Copernicus to dispel this notion. Not only was the earth not the center of the universe, it was not even the center of the solar system.

This discovery gave rise to the Copernicus Principle, which states that the earth is positioned where it is by happenstance, and so there could be literally thousands of earths in our galaxy and universe that can sustain life. For decades this way of thinking has dominated the scientific world until a group of scientists began to think about what it takes for a planet to sustain life. As these scientists piled up facts and evidence of the highly complex and finely tuned requirements to sustain life, they came to the conclusion that not only is earth not haphazardly positioned, but it is precisely where it needs to be for life to exist on it. This is called the Anthropic Principle, which means that earth was specifically designed by someone to support humans as a life form.

22

Why is there such a radical difference between these two positions (besides, of course the obvious conflict between those who acknowledge the existence of God and those who don't)? The answer is that both the medieval church and the naturalist have confused centrality with significance. That is to

say, that all things of primary importance, like the cream center in an Oreo cookie, must be at the center. Just the opposite is usually true.

Allow me to use a few illustrations. There are many things that I do not want to be the center of. For example, I do not want to be in the center of a storm, the center of a raging fire, or the center of a tense controversy. The list of central situations to be avoided is very long, indeed. I do not want to be in the center of a bomb strike, the center of an epidemic, the center of a flood, the center of a vat of hot asphalt, or the center of a criminal investigation. Everyone knows that mom told you as a young pup to stay out of the center of the road. Enough said.

The same can be said of our beloved earth. If, for instance, our earth was in the center of our galaxy, it could not sustain life. The radiation, energy levels, rate of explosive star formations and novas, and the magnetic fields are just far too strong to support life. No, earth needs to be tucked away in a quiet corner of the galaxy where there are few stars (just enough to provide the adequate amount of energy for us) and limited stellar activity. Guess what? That is exactly where we are. Suspended between the Perseus and Sagittarius arms of the Milky Way (about a third way up the arm), we are in a position of few stars (this is why the sky is not covered with tens of thousands of visible stars) and are not bombarded with massive radiation and matter.

Where earth is located is a wonderful location to be in. We can look out of our galaxy to see other galaxies (Andromeda, M-33, and others) and

23

down the arm of our own galaxy for a peek at the center of the Milky Way (the Beehive).

But wait; there is more good news for our marvelously located planet. We happen to be located at the exact location we need to be in our solar system as well. To sustain life earth needed plenty of liquid water (water is liquid in a very narrow temperature range—our closest neighbors are Mars and Venus—the first is frozen & the latter is hotter than liquid lead), a molten iron core to produce a protective magnetic field to shield us against harmful radiation from the sun, a transparent atmosphere to allow sunlight to filter through it while also allowing heat to escape, a right sized moon to help stabilize our orbit without freezing it in place (as the moon's orbit is), an abundance of minerals, and just the right amount of gravity. Besides all of these things, the earth had to be tilted just right (23°) so as to keep the planet the correct temperature (direct sunlight would burn up the middle of our planet and freeze much of the north and south), we needed a yellow G class star to give us just the right amount and type of light. and we needed the precise strength of gravity for our bodies to function. This list goes on and on.

The area of our solar system in which life could be sustained is called the Circumstellar Habitable Zone (CHZ), and we happened to be snugly inside this narrow band. Five percent farther away or twenty percent closer and we would be out of the zone completely. Not only so, but our orbit is slightly elliptical, whereby we are farther from the sun in the summer and closer in the winter. Since the vast majority of the land surface on earth is in the north hemisphere, this works just fine.

Once again this proves the old real estate adage--it's about location, location, location. We are pleased, then, not to be in the center of a congested crime filled city, but instead in a nice rural outer suburb. It is here that we thrive, and nowhere else. Indeed, our happy situation is so finely tuned that

24

another earth could not occur by happenstance in a million universes. We have been specifically placed in our wonderful home by the Master Realtor, who knew just what we needed.

Chapter 6
Fossils on a date and seashells on river banks (the Earth's age)

I married an outdoor loving wife. She is athletic and has kept me mobile and fit through more than four decades of marriage. We often hike and kayak when we are able to do so. One of my wife's favorite hiking trails is the Billy Goat Trail near Great Falls in the C & O Canal National Park in Maryland, north of Washington, DC. Over a two hour trek we climb up a trail over boulders to a rocky overlook that peers down on the Potomac River and across into Virginia.

As we descend down toward the river bank we finally come to a little beach next to the river. Now, mind you, this beach is above tidal water. That is to say, the beach is not on sea level, so the water is fresh water and not ocean water. Yet, even though there is no ocean water washing up on this little beach, the beach is strewn with thousands of seashells. Where did thousands of seashells come from when the water that carried them is fresh water? They are washed out of the mountains. In the case of the Potomac River seashells, they are washed out of the Catoctin, Allegheny, and Shenandoah Mountain ranges. Most other rivers draining out of mountains will have seashells on their banks as well.

What's with seashells coming out of mountain chains? There is only one explanation for this. At one time all these mountains were under the ocean. In fact, all the land on earth was at one time under the ocean. For this reason giant fossils of fish have been found in Kansas (definitely not ocean front property) and ancient ocean fossils have been found in Colorado.

Now, let's think for a minute. Where have we heard of water covering the entire face of the earth? Oh yes, I know. The Bible tells us that there was at one time a universal flood that covered the entire earth. We call it Noah's

Flood (Genesis 6-9). As a matter of fact, all fossils have been laid down by a massive burying of material carried by water.

This contradicts the prevailing opinion of naturalists. The evolutionist describes a slow process whereby fossil layers are built up over time. Such could not have occurred. If you bury an animal in a shallow grave, it decays and turns to dust. Only when the animal is rapidly buried so that oxygen, microbes, and insects cannot get to the carcass, will a fossil be produced. If too much material is buried over the top of much vegetation or fauna, then oil, coal, and gas are produced. At times the coal is compressed to graphite and then diamonds (by the way it does not take millions of years to make diamonds, for they make them in labs today and they are called zirconium).

Obviously, evolutionists will take exception to what I have just written, but they cannot overcome common sense. If your pet hamster dies and you bury it in the back yard, a few months later there will be nothing left of Hammie. His body will have turned to dust. All the epochs and ages laid out by naturalists for the fossil column are made up out of their own imaginations, then. The millions of years of fossil production never occurred.

Let's just take a few examples to demonstrate how the fossil layers could not represent millions of years of life on earth. First of all, we are told that life evolved over many millions of years. Evolution needs time plus chance plus matter to work. Of course, while the biology teacher is teaching a slow development of life over a long period of time, next door the physics teacher is telling his students that things break down over time (entropy = matter goes from an organized to a disorganized state over time--just think of a car growing old and rusting away and you will get the idea).

If this process is so slow, then we would see slow development in the fossil record and many transitional forms between one animal and the other. We see neither one of these. To the first point, archeologists have discovered

27

a Cambrian explosion. That is, at a certain level in the geological stratum there are many fossils of many different animals suddenly appearing on the scene. There was no slow development, but fully formed complete animals of many kinds. For this reason evolutionist Steven Jay Gould stated that the fossil column is not a friend of the evolutionist. Toward the end of his life he tried to claim that evolutionists would expect an equilibrium and stabilization of species in the development of life at some point. Of course, this is false, since no random mindless process will ever produce a complete equilibrium.

Another problem is the fact that the fossil bed just will not fully cooperate. For example, the Grand Canyon fossils are upside down to what the evolutionists expected them to be. This inversion is explained away by upheavals of the land, but such an explanation just will not fly.

One last thing I need to mention before we deal with Noah's Flood or The Great Deluge in detail. We need to deal with a little chemistry here. All organic life is carbon based. When a plant or an animal is alive, it continually absorbs Carbon-14. This is a radioactive form of carbon. Over time C-14 decays to its non-radioactive state, C-12. The half life of C-14 is approximately 5500 years. This means that in 5500 years one half of C-14 will have decayed to C-12. In another 5500 years half of the remaining sample will decay to C-12. This process continues until there is no more C-14 left in the sample. It would only take about 30,000 years for this to happen. Since all fossils have traces of C-14 in them, then these fossils are less than 30,000 years old. They could not be millions of years old. The chemistry just will not allow it to be so.

What do scientists do about the C-14 problem and other problems like that shallow dust layer on the moon or erosion rate issues? Well, there are two basic responses to these dilemmas. The first is the "aha!" moment (sometimes called the "eureka" or "I found it" moment) where scientists are forced to rethink their positions. They revisit their premise and either modify it or discard it all

28

together in order to form a new better premise that aligns with the facts. This is why numerous scientists became Intelligent Design proponents. The Darwinian model just did not work.

The other reaction is the "huh?, this just can't be" response. These scientists cannot believe that their premises can be wrong. In spite of mounting evidence to the contrary, these scientists hold onto their beliefs no matter what hard facts oppose them. If geologists find footprints of dinosaurs and man in the same rock bed in a river in Texas (see the film *Footprints in Stone*), they will deny that such a thing is possible. If they see C-14 in fossils that they believe should have none, they puzzle over it, contrive new explanations to explain it away, or merely shrug and move on. You may think that this is somewhat dishonest, but strongly held belief systems die hard. Unfortunately, when a premise or preconceived notion is set in concrete, science and our students suffer for it.

Now for a brief word about Noah's Flood. According to the Bible the earth was made with waters above and waters below (on the earth). Genesis 1:1-2 describe the waters above (the Hebrew word for heaven means waters above). This atmospheric ocean produced a cosmic radiation shield which allowed man and animals to live an extremely long time. Dinosaurs and reptiles grew to gigantic proportions (reptiles grow all the days of their lives, shedding skin or adding scales as they grow, so if a reptile lived to be 300 years old, it would be large indeed). Archeologists have found 45 foot long crocodile fossils and 60 foot long snake fossils (compared to 20 feet today).

The earth was a natural terrarium with a mist watering the earth instead of rain (Gen. 2:5-6) The entire planet was lush and green and the growing season was year round. Archeologists have found tropical plant fossils in Antarctica. There was no need to eat anything but plant based food until after the flood when the earth entered the grip of extreme weather

29

conditions (Gen. 9).

The Flood itself produced a massive burying of material. The animals that were more mobile lasted the longest, but even they were eventually overcome with water. Several things transpired after the water canopy was lost. First of all, the earth underwent radical temperature changes with snow and ice in some areas and desert in others. Secondly, the lifespan of all life, including man, dropped rapidly from several hundred years for many animals to under a century for man, reptiles, and mammals. Only the great redwoods and bristle cone pines, which were seeded in right after the flood, reach ages of thousands of years. Only the great whales and sea turtles, who have the radiation shielding protection of the ocean live to be hundreds of years old.

All fossils were laid down by this great flood. We see, for example, some thirty-six layers of trees piled up in a cliff side in Yellowstone, as the flood washed these massive forests up against the cliff side. The pressure of the flood carved the Grand Canyon in a few hours (the little Colorado River could not have done so--consider that the Amazon River drains more water than all the rivers of North America combined and has not carved a canyon).

One more thought before we leave the subject of floods and fossils: hydraulic pressure under the earth would have squeezed all the water and oil through the rocks of the earth if the earth were more than a few thousand years old. Therefore, the assumption that the fossil bed represents millions of years is just naturalistic mythology.

Chapter 7
Dating methods and core samples

I was watching a series of videos from a Christian biology teacher that someone lent to me. I will leave this man nameless since his science is very good but his political views are somewhat controversial (since I am not dealing with politics in this volume, I do not want to get us sidetracked). One of the issues that this scientist was dealing with was dating methods. There are three assumptions that scientists make in using various dating methods. First, they assume a fixed starting point. Second, they assume a constant rate of either progression or decay. Third, they assume that anomalies in their findings have a reasonable explanation that will verify assumptions one and two in the end.

Before I go on, allow me to deal with the last point first. Using the half-life of carbon, scientists have dated ancient artifacts. Of all the radioactive dating methods, this one should be the most accurate since it deals with the shortest time span. Yet on more than one occasion, the method has proven to be horribly flawed. For example, a forty year old manuscript was declared to be 1300 years old by this method. Other examples can be cited, but let's move on.

One thing that bothered the Christian biologist in the film was the fact that core samples were taken from the Arctic that covered 130,000 years (at least, that was how many rings or layers the core had in it). He knew that this could not be the case since he knew that the earth was much younger than that. Then he ran across a man who was part of the team that uncovered a lost P-38 plane in Greenland and restored it. He asked this man several questions about the recovery job.

What the biologist found was astonishing. First of all, in just four decades the plane was already under fifty feet of snow and ice. Secondly,

31

there were ten to twelve rings formed in the snow every day, depending on the angle of the sun and the melt and freeze cycles. These rings were not annual tree rings. Not one ring formed every year, but some 3000 formed in a year. So the core sample stored in Colorado represents a mere four centuries of ice and not 130 centuries.

The same is true of tidelet core samples. Supposedly, there are tidelet core samples that represent 800,000 years, with each layer representing a year. The only problem is that oceans do not have annual cycles. I asked my class what cycles do oceans go through. They all knew the answer. Oceans go through tidal cycles twice a day (even the name tidelet core samples give us the hint of this). So these 800,000 rings represent a mere thousand years and not eight hundred centuries. You can even do your own experiment to prove it to yourself. If you plant a bucket in the sand at the shore where the tide comes in (and if it is not washed out to sea or totally buried) and come back in two days, you will have four distinct layers of sand in your bucket.

This same problem plagues all dating methods. For example, the half life of uranium is 4.4 billion years (U-238 decays to U-235, which further decays to U-232, and finally to lead). By measuring the amount of uranium and lead in a sample, one can determine the age of the sample. Of course, that is to say that we could do so if we knew that the sample did not already start out as part uranium and part lead. Since very few things in nature are formed purely of one material, then this assumption cannot be made. Rocks recently spewed out of volcanoes show a mixture of material. So without even considering decay rates, this method is just not useful.

To demonstrate how useless the method is, many scientists have pointed out that helium is released into the atmosphere during radioactive decay. There should be much more helium in our atmosphere if the decay of radioactive isotopes has been going on as long as scientists claim that it has.

Of course, the dating method proponents point out that helium escapes the atmosphere. Unfortunately for them, the sun adds an equal amount of helium to our atmosphere as is being lost.

I have asked scientists how they can assume the makeup of the initial starting point of their radioactive material, and I was told that they must assume this starting point in order to use the method. When asked what would happen if their assumption were wrong, they merely dismissed the idea as preposterous. This is circular reasoning. It is like saying that a sweater must fit me because I need it to, even if it is two sizes too small. Wishful thinking is not science.

Chapter 8
Mineral soup (is it the stuff life is made of?) and I am not half the snake I used to be (transitional forms)

The same scientist that I mentioned in the previous chapter was speaking before a group of biology teachers. He mentioned that they all believed that they came from rocks. One teacher protested vehemently that she did not believe that she descended from rocks. He asked her how she thought that life came about. She said that there was a primordial pond full of minerals that were supercharged by lightning. From this charging simple amino acids began to form, which eventually led to proteins, the building blocks of life. When asked where the minerals had come from, she then realized that she believed that the minerals were washed out of the rocks. Ergo, she believed that she was a descendent of a rock (this has profound implications on many levels and explains much).

The primordial soup theory is very common among developmental biologists. Unfortunately for them, the theory is flawed through and through. Let me just list a few problems. One, amino acids are not formed by passing electrical charges through minerals. After many experiments in a laboratory under very controlled conditions, scientists have not been able to produce amino acids (let alone proteins). Two, if an amino acid could be formed in this fashion, it would have to wait around hundreds of years to take the next step of bumping into other amino acids to form proteins. Mr. Amino Acid just could not wait around that long. Three, amino acids and proteins are not formed this way. In fact, proteins are formed in a very specific step by step way within a cell with complicated machinery that forms, shapes, and places them in specific structures with other proteins.

The idea that somehow a lot of minerals bumping around in water and being struck by lightning will produce the building blocks of life is just wishful thinking. Let me use an illustration that I used in *The Debate*, second book of The Danny Carter Series. Legos came after my time as a child, but we had Lincoln Logs, brick sets, Tinker Toys, and Electra Sets. However, my boys (we had five of them and no girls, but thankfully, we have granddaughters) all had Legos and K'nex sets.

Now, no matter how many times my boys tossed the Legos onto the floor, no structure formed from the mess that resulted. They could kick the pieces, throw them, roll them, slide them, or fling them, but no recognizable structure would form. For a model building, ship, plane, or car to be produced, they had to specifically put each piece in place, interlocking the individual blocks in precise patterns, for form and function to take place. They had to be the intelligent designers to make the designs. How much more would the incredible complexity of proteins need an intelligent designer?

Even though we cannot get a chicken out our primordial soup (it happens the other way around), let's assume that we have an animal already formed. Let's take a snake. Many biologists believe that reptiles were precursors to birds. Now, let's say our snake slowly started to grow wings (I am not sure what would prompt it to do so, but let's move on).

One of the Darwinian principles is the survival of the fittest. Those animals that have a competitive advantage (speed, strength, size, etc.) over other animals, will more than likely survive when others are eaten. It is the slowest gazelle that gets caught by the cheetah. Since this is the case, a snake with two bumps coming out of it would be at a competitive disadvantage and would not survive in its environment. In fact, all transitional states are at a competitive disadvantage. Fish with bumps on the way to become legs, gills on the way to becoming lungs, or scales on the way to becoming skin have

35

would be at such a disadvantage in its environment that they would become easy prey.

This problem is so impossible to overcome that Stephen J. Gould suggested the hopeful monster theory. As absurd as it sounds, he suggested that a snake laid an egg and lo and behold, a bird chick popped out. This theory is called macroevolution and it is as ridiculous as it sounds. Obviously, snakes beget snakes and not birds. Such changes cannot and do not occur. Also, not only would you have to have two birds (one male and one female) pop out, you would have to have a mother bird to take care of the two chicks, for snakes do not know how to do so. Oh, and by the way, snakes eat birds, so this arrangement would produce a very dysfunctional family, indeed.

Genesis 1 and 2 tell us that everything breeds after its own kind, and so it does. Cattle produce calves, chimps produce other chimps, and athletes produce little athletes (at least, they hope that they do). No hopeful monster theory can rescue the transition problem and no mineral soup can produce amino acids and protein chains.

Chapter 9
A field trip through Cell City

One of the popular stories written by Dr. Seuss is *Horton Hears a Who*. If you remember the story, Whoville was located on a speck of a flower. The people in Whoville were unaware that they were so small compared to our world that no one here could hear them there. At the end of the story, at least as one version of the story tells it, the mayor of Whoville hears a small cry for help from a people on a speck in his world—a people so small that they could not be seen at all.

Now let's call these people the Nanites. The Nanites are so tiny that they are smaller than the width of a cell wall in our human bodies. Why, a human hair would be the size of a whole redwood forest to them. However, let's also assume that these Nanites are aware of us or at least aware of our cells and the micro organisms in our world.

On this particular day at Nanite High School for the Sciences, the students in Mrs. Microplasma's class are preparing for a field trip to Cell City. "All right students, we are going to board the immunocapsule that will protect us from antibodies that would otherwise attack us and destroy us. The capsule's special shield will prevent the host's immune system from recognizing us as an invader.

The students and Mrs. Microplasma enter the capsule along with the driver, Golgi. The host they chose for the trip is old Joe. Since he sleeps a lot, he could be depended upon not to move too far. He would be completely unaware of their presence. Golgi flew the capsule to Joe's ear and they entered a pore and then a cell through its semi permeable membrane.

"Now students take notes as we go. The clear transparent sides of the capsule will give you a good look at the inside of the cell."

37

Terry was in awe. "Ooh. What is that big thing over there?"

"That is mitochondria."

"What does it do?"

"It provides energy to the cell and uses some of the energy from proteins in the cell for its own needs. It also will destroy the cell if the cell begins to malfunction. Some scientists believe that since the mitochondria has a nucleoid, which is not as well developed as a full nucleus, that at one time it functioned on its own as a primitive cell."

Johnny raised his hand and was recognized by the teacher. "If the mitochondria provides energy to the cell and needs energy from the cell to function, how did it function before being part of a cell?"

Mrs. Microplasma was stumped. "I am not sure."

"Do we see any mitochondria floating around by itself today?"

"Uh, no."

"Then what is the basis for saying that it must have been an independent cell at one time?"

"Uh, well...we better move on now. Oh look there is the Golgi apparatus. It helps to package proteins, particularly ones for secretion, so it helps handle waste products in the cell. And look over there. That is the nucleus which carries the DNA code and chromosomes and functions to direct activity in the cell and to lead in the function of dividing the cell."

The students saw machines that shaped proteins, transported proteins, opened and closed gates in the cell, and moved food and water and oxygen into the cell, and waste and carbon dioxide out of the cell. They watched as cilia precisely moved material around the cell through turning on and off each hairlike protein at precise moments. They learned of cytoplasm, ribosomes, lysosomes, and flagellum.

They were particularly amazed at the DNA double helix structure with

38

its hundreds of thousands specifically placed proteins. They watched as the RNA duplicated the DNA in order to make exact copies of the DNA. There was just so much to see that they could not take it all in in one field trip.

Toward the end of the trip, the driver, Golgi, detected a build up of energy in the center of the nucleus. "Mrs. Microplasma, I believe that the cell is beginning to build up energy for duplication. We better leave before it does."

"Okay, take us out then. Class, as we leave, are there any questions?"

"Mrs. Mircroplasma, don't some scientists believe that Cell City slowly developed?"

"Yes, Mary. They believe that the cell evolved, moving from simple to complex."

"What was the cell before it was a cell?"

"I am not sure."

"Well, does not the cell need all the functions and parts that we saw today to function? What could be taken out and it still be a functioning living cell?"

"I am not sure. It needs everything that we see here to function."

"There has to be a way to get food and water, to remove waste, to ward off attacks, to duplicate itself, to make proteins, shaping and placing them in the cell, a way to repair itself, and a way to transport matter in, around, and to the outside of the cell. How could this complexity and function come about by some slow process?"

"Now that you mention it, I suppose that it could not do so. All its parts have to be in place for it the cell to function. I guess there had to be a designer that made such a marvel as this."

Mrs. Microplasma is correct. Cell City had to be specifically and carefully designed by a great engineer. All the parts of the cell have to be in place for the cell to function. This is called irreducible complexity. Let me illustrate it this way. When you get into your car, buckle up, make sure your cell phone is off so you won't be tempted to answer it while you are driving, and then turn the key in the ignition switch, several things happen (assuming the car is functioning properly). A spark from the battery sends electricity to the starter motor and turns the motor, which in turn rotates a gear to start the engine. A spark ignites gasoline in the piston chamber and the resultant controlled explosion provides energy to move the piston up and down. The piston then in turn moves the crankshaft, which rotates the drive shaft, causing the car to move when the car is put in gear.

For this car to function, it has to have the engine, starter motor, battery, transmission, drive train, wheels, and a myriad of other parts. Each functioning part is necessary for the car to work. If the battery is dead, the car does not start. If the starter motor is broken, the car does not start. If the transmission is broken, the car does not move, even though it will start. All these parts must be present and in working order so that the car will move.

No car manufacturer sends out cars with missing parts, telling the customer that they will gradually send the needed parts over time. They would sell no cars this way. In like fashion, a cell could not have evolved by slow random processes. All the cell parts are necessary for the cell to function. Any part that is missing would make the cell nothing but a dead collection of matter.

Since the cell that Mrs. Microplasma's class visited is far more complicated than a car, computer, or anything else that man has devised, it surely had a great creator. For this reason David declared that we are fearfully and wonderfully made (Psalm 139). So the next time someone tells you that the cell evolved, have them show you how each part could have come about separately from the other parts and what those parts were doing while they were hanging around for the other parts to arrive. Since each part depends on

40

every other part to live, then there is no mechanism to arrive at any one part without it being a part of the whole.

Chapter 10
Micro outboard motors, hot gas spraying beetles, giraffe necks, and other unique biological structures

One of my wife's and my favorite vacation spots is Yellowstone National Park. The park is unique. There is nothing like it on the planet. We would marvel at the geyser fields, mineral springs, smoking caves, paint pots, fire river, petrified forests, layers of fossilized trees embedded in cliffs, a large lake, and massive mountains. The uniqueness of this volcanic field makes it a travel destination for millions from around the world.

Uniqueness in the geography and geology of the earth is a function of tectonic conditions under the earth. However, uniqueness in the plant and animal kingdom is a function of special creation. One of the favorite ploys of evolution is to line up pictures of animals in a sequence from simple to complex. They then pronounce the decree that this shows the progression of evolution. Of course, this no more shows the progression of evolution than lining up a lawn mower, motorcycle, car, van, and truck and declaring that this showed an evolution of transportation, where one vehicle came from the preceding one. God made life on all possible levels and scales. Each life form is unique in its own right.

For evolution to work there has to be many transitional forms. None have ever been found. However, as if to make sure any concept of a progression is confounded, God has placed some very unique creatures on this planet with no possible connection to any other creature. For example, almost all animals have blood based upon iron, but the horseshoe crabs have blood based upon copper.

The list of unique forms in nature is extensive. The neck of the giraffe, the platypus, the defense mechanism of the bombardier beetle, the

42

desalination glands of the seagull, the Venus fly trap and pitcher plant, and the elephant's trunk are all unique structures.

There are no transition forms and no close relatives to any of these animals. They stand alone in nature as unique and wondrous in their form and function. It is odd, however, to hear some evolutionists speak of these unique designs. Often they will say something about the Venus fly trap responding to generations of flies landing on its pedals or the giraffe needing his long neck to battle other male giraffes in combat over females, as if random mindless processes had a mind or as if a plant or animal passes on to succeeding generations the need to catch flies or battle other males. Such a notion is ridiculous. When a flower dies, it is dead and transfers no memory in its seeds to other flowers. When animals that use horns, strength, teeth, or claws to do battle, they do not pass on to their offspring the idea that somehow a long neck would be helpful to use in combat.

Let's talk about the unique engineering of the giraffe's six foot long neck. The giraffe's long neck is due to an exceptional elongation of the vertebrae in the giraffe's neck. Each vertebrae is approximately eleven inches long and the vertebrae make up an amazing 52% of the length of a giraffe's neck, as compared to 30% in other grazing animals. Not only is this true, but the giraffe has a unique system in place to allow it to take a drink. You see, if the giraffe would bend down to take a drink it would faint as the blood rushed down its long neck to its head. But, of course, the giraffe doesn't faint because this amazing animal was designed with special valves in its neck that shut off blood flow to the head so the animal could take a drink. There is no chance at all that such a system could come about by slow random processes. This marvelous piece of engineering was designed by a great creator.

Another unique design is the desalination glands in the seagull. In the *Rime of the Ancient Mariner* by Samuel Taylor Coleridge, the famous line

43

"water, water everywhere and the boards did shrink; water, water everywhere, but not a drop to drink" indicates the problem that mariners have faced for centuries. You just can't drink saltwater, for you will first go crazy and then die if you do. Why is this? It is because of a process called reverse osmosis. Saltwater draws moisture out of cells rather than putting moisture into cells. Seagulls do not have this problem. They have special glands in their heads that filter out the salt from saltwater so that they can drink right out of the ocean and not be harmed. A seagull can just land on the ocean surface and take a drink. Again, there is not any chance that such a system could have developed through slow random processes.

Not much needs to be said about the platypus, for anyone looking at the platypus can see that it is one of a kind. In fact, the animal is such a hybrid that it looks like it was put together by a government committee. The animal is a mammal with a duck-like bill, but with teeth, fur, poisonous spurs on its hind legs, and just for good measure, it lays eggs. Talk about a confused animal! Is it a bird, a beaver, or an alien? Of course, it is none of these. This incredible animal is a uniquely designed highly efficient marvel of a very imaginative creator. There are no other animals from which this wonder of nature has been derived, and no slow random processes could have produced it..

This brings us to the elephant's trunk. Is it a nose? Is it a hand? Is it a hose? Actually, it acts as all three. The elephant's trunk is so strong that it can pick up heavy logs, but so delicate that it can pick up a peanut. The elephant can vacuum up water and then take a drink or shower. This stupendous organ is a testimony to a master engineer and designer.

Let's look at one last creature to demonstrate unique design and creation in the animal kingdom. The bombardier beetle is a fantastically put together critter. Its machinery is a marvel to behold. When a bird, spider, snake, or insect creeps up on this little beetle, its defense mechanism is

44

triggered. There are two holding tanks in the abdomen of the beetle. One holds hydrogen peroxide and the other hydroquinone. The mechanism the beetle uses to make these chemicals is a story in itself, but let's continue. The little creature quickly sends these two chemicals down tubes into a mixing chamber where a catalyst is added to the mixture (the catalyst is made in a little chemical factory in the insect). The mixture quickly heats up and then the beetle sends the hot gas that is produced down a tube to be injected out the back end of the insect at 220°F, thus discouraging any would be predator from taking a bite. Every single part of this mechanism must be in place for it to work. There is no conceivable way that a step by step slow process could bring about such a device. Chemicals without holding tanks would kill the beetle. Tubes without tanks and chemicals would serve no purpose. Tubes, tanks, and chemicals without ejector muscles would put a boiling soup of hot gas stuck in the beetle. Like the building of a great machine, every part of this process was designed, planned, engineered, and meticulously built by an incredible creator.

Before we conclude this chapter, let's look at one pretty awesome plant--the Venus flytrap. In the old Alfred Hitchcock TV show, Hitchcock would open the show with his famous big belly filling a silhouette drawing on the wall. However, at the end of the show he is swallowed by a giant Venus fly trap. Only his legs are sticking out. As a child, I actually thought that there were such man eating plants. Of course, this was nonsense, but what was not nonsense was the fact that there is a small plant living in the swampy areas of the southeastern part of the United States that is carnivorous. This plant eats ants, spiders, beetles, and flies (but not people).

Now if you would look up the article on the Venus fly trap on Wikipedia, you will discover that the writer of the article proclaims that the Venus fly trap evolved to provide for itself nutrients in a nutrient poor

45

environment. There is nowhere in the article that lays out a detailed process of how this could have happened, because there are none. First of all, a plant that cannot handle a nutrient poor environment would not be able to grow in such an environment, let alone evolve into a carnivorous plant. Secondly, the sensory mechanisms and protein trigger devices in this plant are enormously complex. Again, no slow, incremental process could even begin to develop such a finely tuned complicated mechanism--not in a gazillion years.

We can cite hundreds of examples of such complex unique structures, but the few examples already covered will be sufficient to illustrate the point. The truth is that even the simplest biological mechanisms must be specifically and meticulously engineered.

Chapter 11
DNA and crystal riding molecules

Professor Dean H. Kenyon carefully considered the question from his student. Being the famous author of *Biochemical Predestination*, he was certain of his theory, but the question gave him pause. The student asked if we see any mechanisms in place today where life and life processes were produced without the use of DNA. Finally, Professor Kenyon had to admit that he knew of no such processes.

Professor Kenyon's theory, which is still widely accepted today, is that given the nature of chemistry, organic compounds and life itself were predestined to happen merely from molecules bumping around against each other (never mind, where the molecules came from to begin with). These compounds just naturally fit together.

The question that the student asked haunted the respected academic. As he pondered the problem, he came to realize that not only do compounds not form by merely bumping into each other (many take both catalysts and certain energy infusions), there was no path for them to ever do so. The professor had written a whole textbook on the what of his theory and even the how it possibly could be without considering the actual physical laws which exist in nature.

After much reflection, Dr. Kenyon announced to the world that he was abandoning his theory as untenable. It just would not work. Eventually, he became an intelligent design proponent. There just had to be a great intelligent engineer to put together the complex DNA structure seen in each living cell. Bumping around just would not work.

47

Most scientists, however, have not given up on the bumping around theory. When Ben Stein interviewed one scientist for his *Expelled: No Intelligence Allowed* DVD, the scientist thought that it was patently absurd to think of a Noah's ark and a Garden of Eden. Then when asked how he thought how life began, he stated that he believed that molecules were riding on the backs of crystals. Ben Stein was incredulous (though with his deadpan personality, it was hard to tell). He asked the question again. He received the same answer. Molecules were riding on the backs of crystals (and he called Noah's ark absurd?).

My brain conjured up an image of a benzene ring with a cowboy hat trying to tame a crystal. I am not sure how a molecule rides a crystal (we do not see such a thing happening today), nor do I know how molecules riding on crystals would help explain the origins of life. Even if molecules could hitch such a ride, they would never be able to combine with other crystal riding molecules. Chemical reactions just do not happen this way.

Obviously, such conjuring shows a certain level of desperation on the part of the naturalist. Even though no such mechanisms are seen today and no teacher could write chemical formulas on the board based upon such imaginings (let alone reproduce it in a lab), the belief still is put forth. What else are they going to do? If you do not believe in an all wise, all powerful, all intelligent creator, you have to invent something. If your inventions are patently ridiculous and unworkable in the real world, so be it. And if the best you can come up with gives aid and comfort to your opponents (if this is the best they can do, then our faith is in great shape), for it is no threat at all to them, there is nothing else you can do.

In real life, highly complex DNA, made up of thousands of genes, which in turn are made up of tens of thousands of precisely folded and placed proteins are duplicated by RNA to initiate the multiplication of life. The

48

marvelous double helix contains repeated patterns of four gene groups, labeled ACTG. In various arrangements (AATG, CTAG, GACA,...), these patterns make up the whole that is the foundation of all of life.

The process of such duplication is so precise and well ordered that no accidental predestination or crystal riding molecules can explain it. The great engineering feat called life continues on as it always has been, attesting to the glory of its marvelous creator.

Chapter 12
Adapt or die

I have had the great privilege of teaching in Bible Colleges in Moscow and Minsk. The Russian and Belarusian people are tremendous and I loved my time that I spent with them. I asked them how they tolerated the extreme weather. They told me that there was no such thing as bad weather, only bad clothing.

The Russians understood the concept of adapting to the conditions around them instead of complaining about them. Adaptation is the key to survival. If an animal, plant, or man cannot adapt then that entity will not survive.

The same principle applies for the biological mechanisms within living things. There has to be an ability to adapt to one's environment. If there were no adaptation, the common cold would take us out. Evolutionists teach that adaptation is actually evolution. As we will see in the next chapter, what happens in species that adapt to adverse conditions within an environment is not evolution, but devolution. Information is not added, but it is subtracted or rearranged.

God has placed in every species (called kinds in the Bible) certain traits. Animals can see, move, hear, sense, think, control their temperatures, and perform a myriad of other tasks. Two of the tasks that are integral to any species are the ability to defend itself against invading pathogens and the ability to heal itself when injured.

Of the many major systems in the human body (skeletal, circulatory, skin, muscle, nerve, reproductive, endocrine, digestive, respiratory, and immune), the immune system is the one that is necessary to defend the organism against disease. It is the job of the immune system to recognize invaders in the body and then to provide a defense against those invaders.

When an invader enters the body, antibodies recognize the invaders and sends out a signal that a defense needs to be mounted. Millions of antibodies go into action. These antibodies hurl themselves at the invader until the right antibody is found. How is the right antibody found? To understand the answer to this question we must talk about shape space. Each pathogen and each antibody has specifically shaped attachment edges. The pathogen uses these ends to grab onto proteins to digest them as food. However, antibodies use their shape space to grab onto pathogens in order to destroy them.

Let's use a simple illustration to provide a clear picture of the concept. When I work on my car, which is rare since I am not good at such things, or around the house (ditto), I have a set of socket wrenches that I use to loosen and tighten bolts. Only one socket matches up with a particular bolt. A 3/8 inch socket will not work on a half inch nut. For example, twice a year I have to take off the positive battery cable on my car battery exactly at noon and then put it back on. Only a certain socket will work to do this (I find it puzzling that the negative cable clamp takes a different size socket than the positive one does). Why do I have to do this? Quite simple: the clock function button on my car radio stopped working. Providentially, I discovered that the radio resets to noon when it loses power. So, twice a year at the time change from daylight savings time to standard time and vice versa, I wait until just before noon and reset my clock manually from the car battery.

In like fashion, only one antibody will match the invading pathogen. The immune system will throw thousands of antibodies at the invader until an exact match is found. When this occurs, the proper antibody drags the pathogen to a T-cell, which in turn sends out a signal to a messenger cell to make more of this exact antibody to fight the invaders. Rapidly, the antibody

51

army is built and the battle is engaged. In most cases, the body is able to get the upper hand and eventually the antibody army heroically overcomes the invaders and not only defeats the threat, but also builds up immunity to any future threats from the same pathogen.

At this point the job is not over yet. With the pathogen defeated, the antibody army must be made to stand down lest it would turn on the body it is supposed to protect (auto immune diseases entail such a tragic occurrence as this). Another protein sends out a signal to start destroying the antibodies, for their job is done. A messenger protein carries the message to make proteins that will destroy the antibodies. However, as the antibodies are being destroyed still another protein is produced to protect the antibodies that are still fighting the remaining pathogens in the bloodstream. The whole process is marvelously engineered.

No evolutionist can explain how such a wonderful system could have developed through random process since the whole system must be in place at the same time for it to work. If the system is not in place and working, then the body dies, for it could not fight off even the most rudimentary of diseases. This process, then, is not evolution, but adaptation.

Adaptation is as necessary to a species as digestion, temperature control, or locomotion. Instead of being a part of an evolutionary system, adaptation is merely an integral part of life itself. No microevolution is taking place, for an animal or plant is not evolving into anything else. Instead, simple adaptation is occurring.

Let me give one more illustration of this. When I was in junior high school, we had an exchange student visit us from Sweden. It was April and in Maryland the temperature at the time was in the 40's. However, the exchange student thought that it was beastly hot. She slept a lot and found the humidity oppressive. Back home where she was from, the temperature was in single digits or below. We were in jackets while she was wearing short sleeves and fanning herself. Her body was adapted to extreme cold and was not used to warmer weather. It would take her many weeks to adjust.

On the opposite end of the spectrum, a few years ago my wife and I went to a pastor's convention in Los Angeles at the end of June. The temperature was in the low 70's and was very pleasant. However, the pastors from LA were all in heavy sweaters. They were used to very warm weather with temperatures around 90 in late June. They were "freezing," much to the bemusement of those from the cooler northern tier of the country. Yet, if LA would stay cooler, even these southern Californians would eventually adapt.

Adaptation, then, is built into the biology of all biological organisms. This quality is not a sign of evolution, but of life itself. Without it, life could not exist.

Chapter 13
Of malaria, fruit flies, and viruses

I admit it. I am a comic freak. I love Peanuts, Poncho, Sherman's Lagoon, For Better or for Worse, Prince Valiant, Dilbert, and a myriad of others. I even like to read Doonesbury, though the extreme liberal slant of the strip does not line up with my political views.

In one of his strips, Gary Trudeau walks through a scenario at a doctor's office. In panel one the doctor informs the patient that he has a certain disease. As the scene unfolds it is revealed that the patient does not believe in evolution. The doctor then asks the patient if the patient wanted medicine for the old form of the disease or the evolved form of the disease.

What Trudeau is promoting here sounds so logical that people readily agree with it. Obviously, pathogens evolve and, therefore, new drugs need to be developed to combat the new forms of these diseases. The problem is, the mutated forms of these diseases are not a product of evolution, but of devolution. Devolution is the process of organisms becoming less organized and vibrant due to the loss of material or the rearrangement of genes.

Let's take malaria as an example. Dr. Behe does a marvelous job explaining what happens in the mutations of malaria. When it was discovered that chloroquine would kill malaria protozoa, science believed that they had discovered how to eradicate this scourge of the earth that had killed millions of humans. For a while it looked like they were correct, but after a short interval malaria made a dramatic comeback. The mutated form was immune to the chloroquine treatment.

Millions of students are taught that this resurgence of malaria was a prime example of evolution at work. Not so fast. Let's take a look as to what really has happened. All the malaria protozoa which could not defend itself

54

died. Think about it. If the protozoa could not fight the treatment, it would cease to exist and thus, could not mutate, let alone evolve.

Then what happened to allow the disease to make a comeback? Actually, the answer is rather simple. There existed a certain small percentage of the malaria population that was already immune to chloroquine. These protozoa did not come into existence or mutate to meet the threat. They already existed. The genetic areas of the malaria that the treatment successfully attacked in the majority population that prevented reproduction just did not work in the minority population.

Why did the chloroquine not work on all of the malaria? The answer lies in a process called biogenetic degeneration. When genes split they often are not duplicated precisely. Some information is dropped, some is misplaced, some is copied twice, and some is only partially transmitted. In humans such genetic defects can cause devastating debilitation, such as mental deficiencies, physical deformities, or incapacity. In malaria the defects helped some of the protozoa to survive man's attempt to wipe out the pathogen.

Since billions of malaria protozoa are produced every year, the chance that some of them would be resistant to treatment is fairly high. After the majority population of the disease was eradicated, the minority form became the majority form and then rapidly multiplied to become the new scourge. Therefore, through a process of devolution the defective pathogen proved to be able to survive when the more complete form was not able to do so.

Let's look at a simple illustration to demonstrate the point. Let's imagine that there were two towns near each other along the same river. We will call the one town Neatsville and the other one Slopston. In Neatsville the town took pride in its village. The buildings were meticulously maintained and the roads and bridges were kept in fine repair.

55

The same could not be said of Slopston. The people of Slopston just did not care about how their town looked or how well the bridges and roads worked. They were fine with navigating potholes and a bridge falling into the river now and again, even though a loss of life resulted through such neglect.

One day a massive army came to attack both towns. In Neatsville the army was able to easily cross the well maintained bridges over the wide river. The town was conquered and ceased to exist as an independent entity.

However, the story was different in Slopston. As the army tried to cross the bridges to conquer the town, all the rickety bridges collapsed, drowning the army. The rest of the army gave up the idea of conquering the town, considering it not worth the price of conquest.

Now, it is obvious that Neatsville was the better run more complete town; but it is just as obvious that the poorer less complete town of Slopstown ended up having a better defense against an attack due to the lack of adequate infrastructure that an attacking army could use against the town. It was not what the town had, but what it lacked that saved it. The same is true of the surviving malaria pathogen. The lack of shape space sites for chloroquine to latch onto saved the defective pathogen forms. Thus those with the defects survived while the more complete forms died off.

Three observations can be made about the malaria pathogens that are helpful in understanding that evolution is not taking place. One, after millions of generations, malaria is still malaria and nothing else. Malaria is not evolving into any other species or kind of life form. Two, malaria still has not overcome some of the simplest obstacles to its expansion. It still cannot duplicate in temperatures below 61°F, and it still cannot overcome Sickle Cell Syndrome. Three, all surviving malaria protozoa are less complete and less organized than the specimens that did not survive. The organism is losing material rather than gaining material. In other words, instead of become better

56

and more complex, the disease is become worse and less organized.

What can be said about malaria can be said in greater measure of viruses. Viruses, the simplest form of quasi-life (many biologists believe that viruses are not living things at all), multiply at a rate a hundred times faster than bacteria. Yet, after billions upon billions of generations, viruses are still viruses and nothing else.

If simple one cell creatures and viruses cannot evolve beyond what they are and always have been in millions of generations, then surely highly complex creatures like apes, dogs, and humans cannot do so in a few thousand generations (let alone a few hundred if the earth is only a few thousand years old). What is not possible for a simple life form (though these forms are still very complex) is way beyond the ability of far more advanced ones.

Allow me to take one more example so clearly illustrated by Dr. Duane Gish in *Evolution: the Fossils say No!* One of the experiments that developmental biologists like to give their grad students is the nuclear bombardment of fruit flies. Since fruit flies create a new generation every three days, they are easy to study. Flies are produced with striped eyes, no legs, no wings, shortened antennae, or misplaced body parts when they are nuked.

After many generations three things can be observed. One, none of the mutations improved the fly. In fact, most changes were detrimental to the species. Two, most of the nuked flies were sterile, so they could not pass onto the next generation any genetic changes at all. Three, after thousands of generations, the fruit fly was still a fruit fly and no closer to becoming anything else than the original complete fly.

What conclusions then can be drawn from malaria, viruses, and fruit flies, then? Again, let's just list three of them. One, genetic material is being lost and misplaced from generation to generation. Two, mutations may help a

57

creature ward off attempts to eradicate it, but only through the use of throwing up log jams and denying treatment places to latch onto through missing structures. Three, all observed species remain the same species as they always have been and nothing else, though many mutations over many generations have transpired.

Chapter 14
Biosphere community organizers: we're all in this together

A navy pilot tells the story about having lunch one day in a restaurant. A young man came up to him and asked if he was a pilot who had served on a certain carrier. The pilot said that he was. Then the young man introduced himself as the pilot's parachute rigger, who worked below deck, making sure that the parachutes were packed properly.

The pilot shook the young man's hand and thanked him. It was at this point that the pilot realized that his life was in the hands of a nineteen year old packing his parachute.

How much of what we depend upon do we take for granted? On a ship there are men who keep the plumbing in repair, keep the engines working, work on the planes, keep the ship clean, and a myriad of other tasks, but they are mostly taken for granted, because in a smoothly run ship, no one questions why things are working well. They just assume that they should do so.

What is true on a ship is also true of our bodies. Few of us wake up and say, "I wonder if my heart is working today?"--or "I wonder if my liver woke up this morning?" No, we do not have to think about the workings of our lungs, liver, pancreas, heart, intestines, immune system, nerves, brain, glands, and hundreds of other systems, for all of these are automatically controlled by the cerebellum in the brain.

This integral functioning of body parts goes down to the cellular level. Like an amazing complicated machine, cells work with each other. The blood carries oxygen, food, and water to other cells and then carries waste away from these cells. Various cells have specific functions that help the entire organism. Some cells fight disease, some are transport systems, some break down food to produce energy, some cells convert gasses into usable forms,

59

and some get rid of wastes. All these complex finely tuned functions work seamlessly and continuously with each other without any conscience effort on our part.

Each of the complex precisely designed components in our bodies must exist all at the same time for us to live. If we are missing the immune system, we die. If we are missing a heart, circulatory system, liver, pancreas, or digestive system, we cannot exist. If our systems do not cooperate with each other, we are in trouble. It is the harmonious inter-working of body systems that makes us function.

Obviously, just like the putting together of a car or computer, the human body and those of animals could not have come about by a slow random mindless process. Life requires a great masterful designer of incredible wisdom, intellect, and power. The body practically shouts that there is a creator.

We do not have to depend upon looking inside us to see the work of the creator. In fact, we do not even have to use a telescope or microscope. Let's once again go back to elementary science class. In this class we learned that spiders and birds eat insects. Some birds eat seeds. All animals breathe air and use oxygen, but plants use carbon dioxide and transpire oxygen. We breathe in oxygen and expire carbon dioxide, while plants do the opposite. Happenstance? I don't think so. The system was designed this way. Both the plant and animal must exist at exactly the same time for either to live.

The same concept is true throughout our world (the biosphere). Spiders could not have evolved millions of years before insects as so many evolutionists assert, because spiders eat insects. Plants could not have evolved millions of years before animals, since they need insects for cross-pollination. The same is true of molds. Being very primitive, molds are said to have evolved millions of years before plants and trees. This does not work,

60

since mold feeds off of plants and animals.

The food source must exist at the same time as the life form that feeds on the food source. Since all life forms feed on other life forms, they all must exist at the same time. This is the circle of life. The balance of nature is only balanced because all components in the biosphere exist at the same time. There could be no random slow processes to bring it about.

Chapter 15
Cascading through the body: every link in the chain counts

The problems faced by evolutionists compound with every new discovery in biological science. The problems are so overwhelming that scientist Robert Jastrow wrote in *God and Astronomers* in 1992 that "For the scientist who has lived by faith in the power of reason, the story ends like a bad dream. He has scaled the mountains of ignorance; he is about to conquer the highest peak; as he pulls himself over the final rock, he is greeted by a band of theologians who have been sitting there for centuries."

The secret fear of many scientists who believe in naturalism and evolution is that the overwhelming evidence uncovered by new discoveries continues to lead away from their belief system and toward purposeful design. From the forty part outboard motor driving the flagellum on the end of a bacteria to the chemical factory inside a cell to the steady state (homeostasis) within a body, the fingerprints of design are everywhere. We will use just one illustration in this chapter--the cascading process of blood clotting.

We have already talked about the marvelous response of the immune system, but the clotting cascade is a different mechanism. We will dispense with all of the specific names of the clotting factors (Stuart factor, Christmas factor,...) to make the flow of the sequence simple.

When you cut yourself, an amazing sequence takes place. If it did not do so, any cut would cause you to bleed to death. Let's walk through the incredible dance that the body goes through to rescue you from death. When you cut yourself on a piece of glass or metal, blood starts flowing out of the body.

If the blood continues to flow out of the body, you are a goner. To prevent this from happening a messenger protein sends out a signal to the nervous system that there is an open bleed in the system. The brain picks up the signal and then sends out a message for another protein to start cutting up fibrogen to make fibrin. The fibrin is rushed to the open wound to make a sticky web-like patch to close up the hole. More and more fibrin is made until the hole is patched and the bleeding stopped.

Now a new problem arises. If fibrin continues to be made, the blood would clot within the circulatory system, causing the system to solidify, killing you. This is not a desirable result, so another messenger protein is employed to tell the body to stop making fibrin.

The brain then sends out a signal to cease making fibrin. Fibrogen is no longer cut up and the system goes back to watchful waiting. You are rescued without having to consciously think about it.

If any of the eight factors that make up clotting (thromboplastin, Vitamin K, serotonin, ...) is missing, clotting is either inhibited or impossible (i.e., bleeder's disease--hemophilia and other problems). The system must work like clockwork or it will not work at all. Each factor and component must be turned on and off in exact sequence. In an amazing dance, the body overcomes the threat of losing blood with efficiency and effectiveness.

For such a system to work (this is true of all cascade systems), all necessary factors in the system must exist and be in place at the same time. It is impossible for this type of system to come about through some slow random process. There is just no path to do so. The system either is or isn't. It can never be just becoming. Like a Rube Goldberg machine*, everything must be precisely positioned for the cascade to work. Anything out of place or missing stops the process.

If evolution cannot accomplish a cascade, it cannot accomplish anything else in the production of life either. Once one factor is impossible for evolution to accomplish, and then evolution ceases to be a viable candidate to

63

explain origins.

----

*Rube Goldberg built elaborate devices whereby marbles rolled down tracks tipping over buckets which released springs and levers in a chain reaction to finally produce a result that could have been done in one motion (The Mouse Trap Game)

Chapter 16
Symbiosis anyone? The story of inseparable pals

Playgrounds are different now than when my wife and I were children. We had metal slides (yes, you could badly burn your legs on a hot day, but if you sat on waxed paper you could go like the wind), merry-go-rounds (yes, usually some smart alec kids would not let you off until you became sick--they thought that it was funny), wooden swings (yes, some kids got bonked on the head running in front of the swings too closely), and seesaws (yes, smartypants kids would jump off their end of the seesaw when it was down and thought that it was hilarious for you to come crashing down to the ground). Today, everything is plastic and kid safe. The seesaw, merry-go-round, and metal slides have all been banished from the play yard.

Let's concentrate on the seesaw for a moment. It was a wonderful piece of play equipment in its day if three factors were present. You could go up and down in a soothing motion and at the top of the ride you could see over the entire play area. The ride worked your legs and arms, as well as practicing balance.

The three factors necessary for a good ride were rare to find, but when they were found, it was worth the wait. One, the two participants had to be close in weight. A 75 pound kid was a bad match to a 50 pound kid. It just did not work well. Two, both kids would have to get into a synchronized rhythm and agree when the ride was over. Three, when the ride is over, the kid who gets off first must hold onto his end of the seesaw and let it down slowly so the kid on the high end was not released to the mercy of gravity.

In nature, there are similar partnerships to the seesaw one. Two entities work together for mutual benefit. When one entity preys on another, the relationship is called parasitic, but when the two entities are in partnership it

65

is called symbiosis. There are many symbiotic relationships that are helpful (like the helpful digestive bacteria in our intestines), but there are some symbiotic relationships that are absolutely necessary for both creatures. Let's briefly look at two of them.

A termite cannot digest wood. Pardon? Yes, you heard correctly. A termite would die if it ate wood if it did not have its symbiotic partner. There is a certain microorganism in the gut of the termite that digests the wood. The byproduct of the digestion of the microorganism is food for the termite, but the termite provides the necessary host environment for the microbe.

Both the termite and microbe must exist at the same time for either life form to exist. One could not evolve separately from the other. Since the termite is by far the more complicated entity, the microbe had to be around millions of years before the termite in the evolutionary scheme of things. Since this is not possible, for the one cannot exist without the other, then these two creatures could not have come about through evolutionary processes.

The second example comes from the farm. Just like the termite, the cow needs help in digesting its food. In the cow's stomachs (they have four of them), there are microbes called methanogens. The methanogens cannot exist without the cow and the cow cannot exist without the methanogens (yes, the byproduct of this combo is methane gas, but that's not a good topic for polite company). Both the half ton animal and the microscopic ones must exist together from their very origin. Again, no gradual processes could have produced this arrangement.

Symbiosis is found throughout the animal and plant worlds. Just think of bees cross pollinating plants (in China a government insect eradication program killed off the honey bee, so now human cross pollinators must be hired to do the job bees once did for free) or the relationship between certain fish and coral. In a symbiotic relationship, both partners must exist at the same

66

time for either one to exist. So once again we see that evolutionary processes do not fit the bill.

Now would be a good time to talk about the would have/could have/should have scenario. Often when faced with these insurmountable problems, naturalists fall back onto the old "given enough time all things are possible" defense. Even when what is being promoted defies all laws of science and probability studies, time heals all naturalistic theories, or so they say.

Another popular approach by naturalists is to make up a fictional story as a plausible path to the impossible. Beyond mythology, these stories are pure fairy tales created out of the imaginations of desperate naturalistic philosophers who insist that their theories are scientific.

If these evolutionists insist on stories rather than science, there is nothing that can be done to help them. However, they should not be believed, they should be held accountable to real science, and others should be warned about them. Even though the work that you are reading is very basic, enough is contained in it to provide the logic and evidence to guard against and oppose the myths of the naturalists.

Chapter 17
Of horses, beaks, and other similar things (does variation mean evolution?)

It is the darling story of Darwinism. The *HMS BEAGLE* anchors at Galapagos Island. Darwin departs the ship to investigate the wildlife on the island. Slowly the light comes on. Darwin categorizes fourteen different species of finches. He notices that they all have different size and shaped beaks.

At this point Darwin and all his devoted followers leap to the conclusion that all these finches must have come from a common ancestor. Over time the finches diversified to take advantage of the various food sources on the island. Some beaks were thick for crushing hard shells surrounding seeds, some were long and thin to take advantage of reaching seeds hidden in crevices of rocks, and some were wide for scooping sand so they could find seeds on the beach. To Darwin, it was obvious that all came from one finch.

Naturalists ever since have run with this concept. When the eohippus fossils were uncovered, the archeologists said, "Aha! Here is the forerunner of the horse." Of course, the deer sized eohippus did not have hooves and only marginally looked like a horse. The animal was probably more closely related to the deer family than the horse family.

It is a good time, then, to talk about leaps of logic. It would be readily acknowledged that weak individuals in nature usually are eliminated from nature (the slowest gazelle, sick cows, and weak lemurs), and that stronger animals live to breed and pass on their genes. However, as in humans where traits in families tend to pass on to succeeding generations, the same is true in the animal kingdom.

Even though some bird families may have bigger beaks, the line never becomes a different species. No one would ever make the claim that the

68

tall light skinned Norwegian was a different species than the short dark skinned bushman. By the same token, the longer beaked honey creeper is still a honey creeper and nothing else.

In our backyard my wife has placed bird feeders. We enjoy watching cardinals, titmice, finches, wrens, nuthatches, woodpeckers, and chickadees visit the feeders. There is, however, one nuthatch that is bigger than all the other nuthatches that visit the feeder. He is perhaps a third bigger than the average bird of his ilk. We call him frankenhatch. When he is on the feeder the other nuthatches stay clear of him. Yet, even for his great size, he is still a nuthatch. We have not noticed any of his offspring reaching his size. He is an oddity and may or may not pass his size advantage on to his descendents. It does not matter in the long run, for big or small, all nuthatches are still nuthatches. Individual variations make no difference in the classification of the species.

Naturalists make the error of classification of kinds through similarity of appearance. In the tech world this has become an issue. When Apple and Samsung were locked in a desperate court battle over patent infringements, billions of dollars were on the line. Jurors had to try to distinguish whether products that looked alike were actually the same. Could Apple patent the rectangle shape with rounded corners design of the tablet, for instance? Could a juror distinguish between two similar looking black boxes that had different electronics inside them? It is a difficult task for a lawyer to convince a lay jury that two devices that look similar are not indeed identical.

There is a saying that goes, "if something looks like a duck, walks like a duck, and quacks like a duck, then it is a duck." Yet, this is not always true. A playground model of a helicopter is not the same as a real helicopter. A camouflaged blow up tank is not the same as a real tank.

I am writing this chapter over the holidays and my wife was carving up

69

something that looked like cantalope. The outside skin was yellowish tan. The inside meat was medium orange. There was no question that what she was carving was cantalope. However, there were three things wrong with this scenario. One, cantalope was not in season. Two, there was no cantalope fragrance emanating from the fruit. Three, unlike cantalope, the item was difficult to carve. The reason why this was the case was that fact that the food item was not a cantalope. It was in fact a butternut squash.

Now a butternut squash and a cantalope are not only two different food items, though both are gourds, they taste radically different. Of course, no one would confuse the two items in a store, for they look different when uncut. But once cut, they are very similar in appearance. The point is obvious. Just because two items may look the same, does not mean that they are related or have a similar ancestry.

A box car and a mobile home may look similar, but one did not come from the other, nor do they have a common root. I may line up animals all day by size, shape, or characteristics, but this does not mean that these animals have common ancestors. The leap of logic that assumes that they do have common ancestors just will not hold up to close scrutiny.

Even Darwin recognized this fact. He stated in the seventh chapter of *The Origin of Species* that if later scientists discovered that the basic cell of the body was more than just a gelatinous glob of goo, then his theory would be doomed--and so it is.

Chapter 18
Monkey business (did we really come from apes?)

My oldest son lives in Idaho. There is much beauty in the state, but the number of tourist destinations is limited. There are the ice caves, Sun Valley, Hell's Canyon, Craters of the Moon, Thousand Springs, Miracle Springs, the Bruneau Dunes, and the Hagerman fossil beds. The Hagerman fossil beds are touted as a great evolutionary study. Of course, since the public is not allowed to actually go down to the fossil beds, no one can check out any claims that archeologists make.

One day my son was helping do technical work for a broadcast of a science program in Boise. The Hagerman fossil beds were the subject of one of the programs. There were students in the audience. Full sized models of the animals of the bed were on display. One fifth grade student asked if the horse model was from a full sized skeleton. The answer was no. The model was extrapolated (projected from a small amount of information--i.e., imagined) from a few fossilized teeth found in the fossil bed.

Huh? Do you mean that an artist took a few teeth and then constructed a model to show us what the entire animal would look like? Precisely! Even though it is nearly impossible to know how the flesh would look like on a fossilized skeleton, these artists imagine (with the prompting of the archeologists who hire them)) what an entire animal would look like from a few teeth.

One of the popular pictures in biology books when I was in high school was a drawing showing a series of pictures of evolutionary development from a monkey through proto-humans to modern man. The article with the picture spoke of Neanderthals and Cro Magnon man. It was all fiction. The drawings were not real. The drawings were made up. Even Dr. Louis

Leakey's supposed discovery of proto-human fossils at Olduvai Gorge (Australiopithicene Africanus) was merely a collection of a few teeth (probably from an extinct simian, ape).

After the genetic code was cracked, with great fanfare developmental biologists announced that the ape and man share 97% of their genetic code. In other words, except for a 3% difference, the ape and man have identical genetic makeups. This is supposed to demonstrate that the ape and man have a common ancestor and that somewhere along the line the lines split. Later, scientists discovered that both the ape and man have three broken genes defective in the exact same place and way.

Laying aside for a moment our demonstrations that evolution is not possible on the cellular, cascade, and biosphere levels, let us look at these claims. First of all, it would not be surprising that apes and man would share much of the same genome, since they both have lungs, hearts, kidneys, thyroids, circulatory systems, skeletal systems, livers, pancreas, digestive systems, reproductive systems, endocrine systems, nervous systems, and immune systems that operate in practically identical manners. There is no reason why the master engineer would invent new systems to do the same functions.

Secondly, the 3% of the genomes that are different represent thousands of genes and this difference makes all the difference in intellect, appearance, function, conscience, ability, and reasoning. It is no minor issue that three thousand or so genes are unique to man. The 3% difference shows that man and ape did not come from a common ancestor.

Next, we come to the impossibility of the common descent scenario. Every one of us have known family members, neighbors, or friends who have had congenital genetic defects. Fragile X, Williams Syndrome, Huntington Disease, Down's Syndrome, and a myriad of other genetic maladies cause

72

much suffering and heartache. Even the slightest variation in chromosomes and genetic composition can cause mental deficiencies, physical weakness, and a shortened life. Since this is the case, can you imagine what would happen if entire chromosomes come up missing?

This is the problem with the ape to man concept. An ape has 24 pairs of chromosomes for a total of 48 and man has 23 chromosomes for a total of 46. There is no way to move from 48 to 46 chromosomes without total disaster breaking out. No species could survive such a transition. You just cannot get there from here (or as they say down South, "that dog won't hunt"). So you can relax. No matter what you think of your spouse's family, neither they nor you descended from apes.

Now would be a good time to talk about a philosophical technique that is commonly employed by naturalists when a major insurmountable problem stands in their way. When ignoring or explaining away the problem does not work, they engage in minimization. Minimization is the technique by which someone treats a problem as minor when the problem is really quite major. It is like a lung cancer patient saying that he has some minor aches and pains or calling the Pacific Ocean a minor obstacle to kayak across.

When a naturalist tries to minimize the problem of chromosome differential between man and ape by making up a story how such a thing could work they are blowing smoke. Dear reader, when you hear a Darwinist try to explain away huge problems like the ones mentioned in this book, you should recognize that he has no idea how to solve such a problem. In truth, he probably knows that there are no solutions to these thorny problems.

73

Chapter 19
Meet Mr. and Mrs. Amoeba...or not (the origin of genders)

For centuries men dreamed of flying. They drew pictures of flying machines and gliders. Leonardo Da Vinci even had a drawing of a helicopter that would actually work. In the eighteenth century men invented balloons to float above the ground. In the nineteenth century men designed, built, and tested gliders. Yet men could not truly fly.

Why could men not fly? The answer is simple. Until the invention of the internal combustion engine, men had no power pack to power the plane. Birds have enough muscle power to fly. Man does not. When Wilbur and Orville Wright put their twelve horsepower engine in their wind tunnel tested airplane design on 17 December 1903 at Kitty Hawk, North Carolina, manned flight was born. It took flight design and a powerful enough engine to make it happen.

Now let's look at Mr. Amoeba, who doesn't desire to fly, but might like a wife one day. Under the naturalist motif, Mr. Amoeba swam around generation after generation doing just fine, but he wanted to be more (actually, in a mindless process, he didn't have any wants, but food and liquid, but let's move on). As the single cell wonder moved up the chain to be a multiple celled creature, he (actually "it", since it had no gender) reached the maximum point of expansion that was possible in its development capabilities. It somehow needed more. What it needed was gender distinction. It had to become two entities--male and female.

How did asexual creatures become bi-sexual? The naturalists have no idea. The male reproductive system is radically different from the female system. Testicles, prostates, and male sex organs are different than ovaries, fallopian tubes, and uteruses. Also, both of these incredibly integrated systems

74

are radically more complex than the more simplistic split and divide methods of simple single cells or the budding process of molds and fungi.

Even when developmental biologists try to imagine a pathway to move from asexual reproduction to bisexual reproduction they strain to come up with even a beginning of such a progression. There are three things that mitigate against such development. One, evolution is a mindless process, so a single cell creature does not sense any need to be anything else than what it already is. Two, any partial development of a reproductive system (let alone two) would not only be useless to the microbe, but fatal as well. Three, both a male and female reproductive system must develop at the same time. This is not possible under a random system, and so it is a show stopper.

Once again, like the thirteenth century man dreaming of flying, there are no paths to get to where he wants to go. It was not until intelligent agents figured out how to make a gasoline powered engine some seven centuries later that sustained manned flight was possible. By the same token, it took the ultimate intelligent agent to produce genders--male and female (Gen. 1:26-28).

Chapter 20
Footprints, buried skulls, monkey hands, and misplaced fossils

When I was in high school I both believed in the Lord and in evolution. There was a popular school of thought back then called theistic evolution. According to this philosophical point of view, evolution occurred, but God made it happened. The position was a compromise between biblical theology and the prevailing views and science of the day.

The logic that led to this compromise went something like this: the first eleven chapters of Genesis were simply a basic outline to explain to primitive man how things came to be in terms that he could easily understand. The more complex explanations were just too complicated for him to understand. Then, so the thinking went, each day in Genesis 1 represented epochs of time and not literal days. Finally, each step along the evolutionary path was engineered by God.

The assumption was made by the theologians who taught this way of thinking that science had proven both that the earth was billions of years of old and that evolution had indeed occurred. Many mainline denominations are still plagued with these misconceptions.

The surrender by many of even the most respected theologians to Darwinism goes back to the turn of the twentieth century when C.I. Scofield promoted the Gap Theory in his Schofield Bible. The Gap Theory stated that there was a pre-Adamic race (before Adam) that rebelled against God and God wiped them out and started over with Adam and Eve. Instead of the earth being without form and empty (Gen. 1:1-2), Schofield taught that the earth became without form and void.

There are many things wrong with this way of thinking, not the least of which is the attack on the truthfulness of the Bible. One, the normal translation

76

of the Hebrew world *bara* is **was** and not **became**. Two, whenever the word day is used with evening and morning as it is in Genesis 1, it always means a twenty-four hour day. Three, in Genesis 1:31 God stated after He finished creation, everything was very good. If there was already a rebellion of a pre-Adamic race, then everything could not have been very good.

Obviously, neither Schofield nor anyone else needed to cave into prevailing scientific opinion, since what the scientists were peddling was not science. Paul warned Timothy to beware of science falsely so called (1 Tim. 6:20). Since the last quarter of the twentieth century, Darwinists have been on the defensive. They have had such a difficult time defending their position because the electron microscope and other scientific investigative tools have uncovered the great complexity of life, that they have long ago stopped accepting debate invitations from creation scientists. They have resorted to the pathetic strategy of treating such invitations as being beneath their dignity (as a professor debating a first grader would be).

Slowly the light bulb came on in my head when I started going to seminars sponsored by creation scientists. I was fascinated as I read Dr. Gish's book on the bombardier beetle. One of the most intriguing films that I watched during this period of time was *Footprints in Stone*. Here before my eyes were footprints of a dinosaur and man in the granite bed of a river in Texas. When they interviewed a geologist on the film, he could not believe that it was true, even though he had seen it with his own eyes.

Time after time, naturalists have run into these contradictions to their theories. They call these anomalies, but anomalies still have to be accounted for. Over and over the tally of out of place items has added up. A human skull in India found at a depth that has been labeled 70,000,000 years old, living trilobites that were supposed to have been extinct millions of years ago, human tools found at levels before humans were supposed to be on earth, and a

77

mixture of prehistoric animals and modern animals in the same fossil bed, all point to a destruction of Darwinist theory.

There is another philosophical concept that should be discussed at this time. It is called Occam's razor. Occam stated that if a proposition is contingent on a continuing lists of contingencies (if this is true and if this is true and if...), then his razor will cut it off. In other words, if a theory is true only if a series of conditions are met, then the theory is not valid. Therefore, if Darwinism requires a continual explaining away of anomalies and an invention of stories to try to explain how intricate designs could have evolved, then Darwinism cannot be true, for Occam's razor cuts it off.

Chapter 21
One shark + one shark = evolution?
(Just what is evolution anyway?)

The year was 2012 and I had just opened my computer. I have my computer set to open up to news so that I can peruse what is happening in the world. On this particular day an article was posted from Australia stating that marine biologists had discovered a shark which was a hybrid of a Northern white fin shark and a Southern white fin shark. Then the biologist who was being interviewed announced that "this is evolution in action." Huh? One kind of shark breeds with another kind of shark and this is evolution in action? Come again?

I asked my biochemist son about the article and he told me that biologists are now defining evolution as any change in alleles (gene groupings). You've got to be kidding. Every time a baby is conceived genes are rearranged. Evolution used to be defined as mutations that lead to changes in species on the way to advancing to become new species.

To define evolution as a change in genes is a tautology. That is, it is saying the exact same thing in different words. Nothing new is being said at all. It is like saying that Rover is a dog, but he really is a canine.

Now would be a good time to talk about taxonomy. The classification of plants and animals into phylum, families, groups, and species are artificial. To say something is a different species really has little meaning. The Bible merely separates life into kinds. A kind is defined as all the animals that can breed to produce offspring (Gen. 1-2). So all dog kinds are one kind. All large cats that can interbreed are one kind. All cattle are one kind. We can distinguish sub-kinds such as wolves, dingoes, foxes, poodles, etc, but all of them are dog kinds. We can identify different cattle types, but they are all

79

cattle kinds. There is no evolution occurring when species that can breed together do so. When two different white fin sharks interbreed, they may have produced a new breed, but not a new kind. No evolution has occurred. Those who claim that it has are desperate to try to keep Darwinism from completely sinking into the abyss.

So what really is Darwinian evolution? One of the basic rules of any game is that you cannot change rules in the middle of the game. This is exactly what evolutionists are trying to do. The basic understanding of evolution is that species continue to change until they become some new species. This is called common descent.

If common descent is true, then we would expect to see all kinds of transitional forms, both in the fossil record and in the world today. Reptiles with partial wings, fish with partial lungs, and snakes with partial legs should be in abundance. We see none of these things anywhere. All species are complete and not moving to any other species. Common descent, therefore, is not true, and the theory of evolution is left with no demonstrable evidence.

Chapter 22
Once upon a time (naturalists tell a story)

My wife and I raised five boys, so we became used to trucks, balls, toy guns, Legos, super heroes, action figures, and martial arts. We had no girls, but we have since had granddaughters. On a visit to my second son's home I took some books with me to read to my three year old granddaughter. When she saw the Golden Book story of *Cinderella* she became excited and wanted me to read that story--over and over again. It was as this point that I discovered that all the rage among little girls now is princess stories. Belle, Ariel, Theona, Jasmine, and Aurora all have their place in the folklore of little girls and these fictional figures are their heroines.

There are certain themes that run through all these stories. There is always a handsome prince or would be prince who is interested in the princess and vice versa. There is always a villain or obstacle inhibiting the lovers from joining. Finally, the prince overcomes all obstacles, wins the princess, and they live happily ever after (this part of the story is never defined and no sequel ever follows showing us what happily ever after looks like).

These stories are full of magic, mystery, and monstrous antagonists. For this reason we know that they are fiction. We know that genies coming out of lamps, magic carpets, apples that produce eternal sleep, kisses that wake princesses from eternal sleep, glass slippers (can you imagine?), and mermaids just do not exist in the real world. Of course, we don't care if they really exist or not. We enjoy the imaginary stories of magical worlds, where the impossible is possible.

There is a difference, however, between imaginary stories that entertain us and imaginary stories put forth as real stories to fool us. Naturalists are adept at doing this. For example, in one interview with Dr.

81

Richard Dawkins, a picture is projected in the background that is supposed to represent an ancestor photo album. The pictures are stacked on end and every now and then we can see part of one of the hundreds of pictures. Starting with a modern human, the pictures going back hundreds of generations begin to look more a more apelike, until the farthest picture is pure ape--King Kong himself.

The depiction is effective, but completely imaginary, having come out of Dawkin's imagination. No process that his illustration projects can happen or has happened. The incredible number of contingencies that are necessary to even begin to bring about a scenario like this one are cut off by Occam's Razor long before any significant changes occur.

Dr. Behe calls the stories of naturalists "just so" stories. Not only are there innumerable conditions and sequences and happenstances that are necessary for these imaginary stories to come true (far more than most of the fairy tales), but since each condition defies the laws of science, they cannot come true without magic or miracles (Carl Sagan speaks of the miracle of universe). Without a designer all these stories must be relegated to the category of fairy tales and nursery rhymes.

Another problem with these just so stories is the fact that for the universe and life to come about in the fashion we see it, there is not only have been a sequence of events that must have occurred, but all the conditions necessary for life must have been present simultaneously (at the same time). The greater and lesser nuclear constants, electromagnet constants, gravitational constants, water coefficient, carbon/oxygen balance, and other factors had to be all precisely set for the universe to work. Just so stories could not have worked at all.

Everything has been purposely set. All the control dials have been precisely positioned according to exact calculations. Neither Cinderella nor

82

Ariel could have ever imagined such a thing. So the next time a professor, teacher, or naturalist tries to give you a just so story, have him show you from actual science how the scenario being presented works in our world in real time. When he cannot show you the hard facts of such a scenario without trying to snowball and bluff his way through it, you will know that another princess story is being spun. Unfortunately, the presenter really believes his fantastic tale and wants you to believe it also.

Chapter 23
Does evolution laugh?

When I was fourteen, my family moved from an apartment to a house. Along with the move, unbeknownst to me, my father threw away some of my prized possessions, including thousands of baseball cards and a neat road map book that showed all the sites of interest along major highways. He figured that I just did not need these things.

Evolution has the opposite problem. There are things that we have that just do not seem to fit into evolutionary must haves. For example, where does a conscience fit into a survival of the fittest motif? How about morality? Where did humor come from? Did evolution need to crack a joke?

How about aesthetics, indulgences, a desire for power, various tastes, and personality? What about cultural differences? Evolution does not deal with any of these things, nor can it explain a need to worship.

Every culture that has not traded its deities for worshiping education and itself has a religious system. How can evolution explain these things? Obviously, it cannot and never will be able to do so.

In one of Oxford Professor John Lennox's debates with Christopher Hitchens, he asked Hitchens how man's universal need to worship a deity or deities fit into the evolutionary scheme of things. To me Hitchen's answer was mind boggling. He stated that man has a servile nature and feels that he needs to submit to something higher than himself. He said that he does not know why this is.

Hitchen's answer was the classic answering a question with another question. To say that man has a servile nature, but I don't know why, is to beg the question. The answer given does not fit the evolutionary motif at all. If evolution is the mechanism that is explained by the survival of the fittest

84

paradigm, then worship does not fit into the picture at all. In fact, the need to worship, instilled in all men, but educated out of many, is a basic human desire. Depression, purposelessness, and cynicism set into many who have no meaningful worship experiences.

The second thing that is baffling about Hitchen's response is that men do not want to be servile at all. Men want to control others. They want power and significance. They want to rise above others. Pride, ego, and desire to be in control consume the bosom of man. So, it is not that men are servile by nature that drives them to want to worship something higher than themselves.

Then what does drive men to want to worship? The answer is rather basic to human nature. Man readily recognizes that he is not in control. Something greater than himself must be in control. Man does not control life or death, storms or disease, the actions of others or even his own emotional responses. He is a captive in a world beyond his grasp. Only in modern societies is there an illusion of control (wealth and technology has masked our basic needs).

The grave rips away this illusion, but instead of squarely facing our own mortality, we quickly bury the dead and move on. Churchill put it this way: "Men often trip over the truth, but then pick themselves up and move on as if nothing had ever happened." I often think of Churchill's quote when I preach a gospel message at a funeral to an audience with plugged up ears and a burning desire to get on with it so they can pretend that nothing has gone awry.

No, evolution has no answer to where worship, fashion, art, humor, conscience, philosophy, music, and reflection come from. For you see, evolution cannot laugh, cry, bow the knee, rejoice, or repent. It has no ability to do so. Time + Matter + Chance can only bring about a survival instinct (though

85

even this is highly doubtful). To speak of self-awareness, a desire to be fulfilled, or a need to seek out one's creator to thank and serve Him are just beyond the capabilities of Darwin's stepchild.

Chapter 24
Biblical time keeper: when 3 days and 3 nights are not 72 hours

When I was in seminary, I met students who had come from all over the world. My classmates came from Japan, Taiwan, Africa, Europe, and Latin America. One particular Nigerian student was having trouble adjusting to American culture. He had asked an American girl out on a date and she had accepted the invitation. He told her that he would pick her up around 5:30 PM. When he arrived sometime after 7:30, his date was not happy. This puzzled him, for in his country the idea of precise time is a foreign concept. The idea of 5:30 PM meant "sometime this evening." To insist on arriving somewhere at exactly 5:30 would be considered rude. In fact, in most cultures, particularly in the East, if one arrived to a house at exactly the time arranged would be considered an insult. The host simply would not be ready for you at that time.

Missionaries have run across this problem on the field. One missionary related that he had arranged with the chief of the tribe to meet with the tribe right after sundown in the village center around a bonfire. The missionary arrived at the pointed time and was ready to speak to the people. He thought that maybe he had not communicated clearly somehow, for no one had come. The missionary sat on a log for a long while, waiting. After about ninety minutes villagers began to straggle in to the village center. They did so for the next hour. As far as they were concerned, they were exactly on time.

In South America, the idea of business appointments is different than in North America as well. In the US, if a man has an appointment at 10:00 AM, that is when he shows up for the appointment, even if he has to wait in the reception area for awhile. If he was to wait a half hour, he considers himself disrespected and ill used. However, typically, in Latin America, everyone who wants to do business with an individual shows up at the beginning of the

87

business day and several business transactions are done at once. No one would consider themselves ill used for waiting all day to finish their business, or even coming back the next day. It's just the way it is done.

The Western world, particularly those cultures extracted from Teutonic northern Europe, is very precise about time. As a matter of course, the Germans and British became obsessive about time. It was an absolute fetish. We see this depicted in the movie *Mary Poppins* where George Banks leaves the bank at a precise moment each day and arrives at home at an exactly planned moment, expecting his wife to greet him and that his slippers, his paper, and his meal would be ready. Admiral Boom then fires his cannon at exactly 6:00 PM. If one's watch was 30 seconds off, it was considered a major failure.

In Western culture time is considered linear. One moves along the time line from one particular moment to the next in marked increments. The better one plans each increment, the more efficient and productive one is. This is considered a good, well organized life.

The Eastern mindset is different. Time is circular, coming around to the same point each day as it did the previous day. The importance of life's sequences is not time and efficiency, but the event and relationships. We can see this in action in many African-American and Latino church services. The service starts when people arrive and ends when there is a collective sense that the event was winding down. To hold such a culture to starting at 11:00 AM and ending at 12:00 PM just would not work. I have had many African-American visitors to my church arrive ten minutes before the end of the service. The services that these visitors were used to would just be ramping up at this point.

What is true about time is also true about telling stories. The Western mind wants preciseness. We want the details in order with exact times

88

attached to the details. The Eastern mind is not this way. Whereas in the West we like time order and preciseness with a strong emphasis on accuracy, but in the East the emphasis is on the story and its meaning. It does not matter if the events are told in proper order or not. It is immaterial if some details are left out or if certain points are emphasized at the expense of other details.

Herein lies the problem. The Bible is an Eastern book steeped in the rich storytelling and time orientation that comes from the Eastern mindset. The man of the Near East would consider it quite unreasonable to pin him down to accurate sequencing and time placement as he tells his story. The Bible is inspired by the Holy Spirit, but it is contextualized within the ancient Eastern societies in which the events are cast.

It is not only unfair, but misguided to insist that Bible accuracy be evaluated by a modern Western rubric*. The Bible is meticulously accurate in all its doctrines and essential details. For example, when Paul explains that God would bless the world through Abraham's seed and not seeds (Gal. 3:16), the point of the teaching depended upon the difference between a singular and plural noun (Gen. 13:15 - note that many translations inaccurately translate seed as descendents or offspring, missing the whole point of Paul's teaching). However, when such meticulous accuracy is not demanded, biblical writers use round numbers or approximations. Thus, one biblical writer states that the children of Israel were in bondage for 400 years, while another states that it was for 430 years. There is no contradiction here since one is an approximation, while the other counts time from the point of entry to the point of exit.

The point of time delineation most criticized in the Bible is the declaration that the Lord was three days and three nights in the bowels of the earth. The Western mind insists on 72 hours, or at the minimum, 60 hours. Yet, that would be considered ludicrous in the Eastern way of thinking about

89

time. Any part of a day was considered a day. The euphemism used was "three days and three nights." As far as a Jew of Jesus' day was concerned any part of Friday, Saturday, and Sunday together were three days and three nights. A part of the day represented the entire day.

*rubric = the direction in which something is heading, a distinguishing feature, or a highlighted fundamental tendency

Chapter 25
Bible accounts: get your facts straight (do differing biblical accounts invalidate the Bible as an divinely inspired work?)

I have lived in the Washington, DC area and have been a Washington sports fan since my family moved into the region when I was five. One of the things that I have always found intriguing is the post game analysis, not by paid talking heads, but by me and my friends. Even though we all saw the same plays, our interpretation of those plays and even what we saw were all different.

I remember one discussion about a championship bowl game where pass interference was called on the last play of the game. I did not think that it was a controversial call at all. The defender was draped all over the receiver before the ball had reached him. Even the picture in the paper the next morning showed the defender's arms wrapped around the receiver while the ball was still over a yard way from him. The call wasn't even close. Yet, there were those who still insisted that the refs were paid off, no matter what the photo showed.

Of course, eye witness reports are one thing, and technical accuracy is something else. We all want our houses built square and strong, our surgeons to cut exactly at the right spot, and our car parts to be precisely calibrated to the right measurements. Yet even here, there are different degrees of accuracy. When I was growing up, we did not have scientific calculators, let alone portable computers (I know, I know--this is ancient history). In fact, you were not allowed to use a calculator in class (calculators back then could add, subtract, multiply, and divide--that's all, folks). For scientific calculations we used a slide rule.

A what? Slide rules, like abacuses, are considered museum pieces today. A slide rule was a ruler with various markings on it and a central rule which was movable along a track. I even had a circular one. It could do basic math functions, logarithms, trig functions, and square roots. We imagined that by accurately setting up a problem and using our highly trained and discerning eyes, we could read between the lines out to four decimals.

Of course, reading to four decimals on a slide rule (particularly the cheap plastic ones sold in drug stores) was nonsense. Yet, real scientists with expensive magnesium slide rules that had really numerous fine lines were accurate enough to build rockets and launch them using these measuring devices. They were pretty good.

However, when scientific calculators entered the market, the slide rule not only became obsolete, but it was also shown to be less accurate than what modern complex equipment called for. During the era when the rule ruled, it was fine. In the era where the computer is king, the rule is considered as primitive and inaccurate as measuring distance by the length of one's foot or the differential between one's palm and elbow (a cubit).

Now, what does all of this have to do with the Bible? Much. In the Eastern mindset, as we have discussed already concerning time, stories are told to illustrate points and descriptions of events are given to include only details that are pertinent to the one telling the story. An Eastern speaker feels no compulsion to line up every event in exact order and at the exact time of occurrence. He considers that what he is telling you is accurate without feeling compelled to be precise. He wants you to understand the essentials of his story so that you can get a broad picture of both the events and the meaning of those events.

The Bible is essentially an Eastern book and does not present itself to be scrutinized by British or German obsessions with measurements and

92

sequencing. So, when Matthew states that Judas left the last supper before the communion was implemented and Luke has him doing so afterward, there is no contradiction between the two, for they were both listing the events, not in time order, but as a catalog of what happened that evening.

The same is true of other events. Three Gospel writers list three different inscriptions on the cross. Since these inscriptions are translations, then leeway can be given. Also, only partial renderings are given in each writing to suit the needs of the authors. Further, one Gospel writer mentions one angel at the tomb, while another lists two. Is there an inaccuracy here? Of course not. The one writer was interested only in the angel that was speaking, while the other wanted to let his reader know that there were two angels present.

The same thing that we have said about events can be said about numbers. Almost all large numbers are rounded off, particularly when dealing with whole tribes or armies. At times, even ages are approximated. If the Bible states that Jesus was thirty years old, He may actually have been 32. The celebration of birthdays was just not an issue then. Even generations were estimated. In the genealogies in the Bible, only major family heads are mentioned. This is why the genealogies do not perfectly align. Every ancestor was considered the father or mother of every descendent. So, the word beget could mean to bear a child or it could simply mean some descendent, no matter how many generations had past. The biblical writers would think that it was strange that you would insist on a complete lengthy genealogy. They wouldn't know what purpose such an accounting would be, and they surely would not waste precious vellum (from animal skins) or parchment and ink producing such a document.

Now, to be completely truthful, there have been some copying errors in the manuscripts over the centuries. After all, we are dealing with 2500 years

93

here. Yet, with all those years of transmission and with governmental attempts to destroy the Scripture, less than 1% of the total documents are impacted (look at your footnotes in the NIV and NKJV Bibles to see these for yourself). Indeed, there are more discrepancies in the seven existing copies of Aristotle's works from the eleventh century than in the manuscripts of the Bible. To demonstrate this amazing truth, the complete work of Isaiah found in the Dead Sea Scrolls was in all practical purpose identical to the copy of Isaiah in Codex Leningrad, some 1000 years later.

What are some of the differences in the texts? Let's list a few. In one account the manuscript reads that 70 men from Beth Horan died, while another account states that 70,000 died when they looked into the ark. Since there would not be 70,000 men in the entire rural region, 70 is the accurate number. However, the difference between 70 and 70,000 in Hebrew is merely a line over the word representing the number. A stray mark changes the reading. Other discrepancies include the missing story of the woman caught in adultery in the Alexandrian texts (Greek texts found in Egypt), the last eight verses missing in Mark in some texts, and the different renderings of how Judas died.

In the case of Judas's death, all of the Gospel writers and Peter in Acts 1 agree that he committed suicide. Yet, one states that he hung himself, while Peter states that he dashed headlong against the rocks and his bowels gushed forth (hopefully, you are not reading this over lunch). Actually, it is not that hard to imagine that both are true. Judas hung himself and when the rope broke, he went down the cliff and dashed against the rocks. Anyone who has ever been to Jerusalem will not have a hard time imagining going over rocky cliffs, for they are everywhere.

Finally, we see global thinking in the quoting of the Old Testament by New Testament writers, even Jesus. Our Bibles are numbered by chapter and verse (put there in the fourteenth century), but theirs were not so. They were

94

on scrolls. At times Paul would say, "somewhere it is written" and no one asked him where, for they were familiar with the quote. At times he would paraphrase an Old Testament writer or quote from the Greek translation of the Old Testament, the Septuagint. He did not feel dishonest in doing so. The quotes were accurate enough.

Jesus would often put two quotes together from two different Hebrew writers. He even said that Jeremiah stated something that Zechariah had actually said. Why? Simple. Jeremiah heads up the prophetic books that included the Minor Prophets (The Twelve) in the Hebrew Scriptures. Often Jeremiah was credited with saying what he had not written due to the fact that this entire section of the Bible was known as the book of Jeremiah.

We do not need to dwell further on this topic, for the point has been made. There are three things I will mention to conclude on this subject. One, all the events and timing in the Bible are in accordance with the Eastern sense of accuracy. Two, the discrepancies due to copying errors are minor and impact no doctrines. Three, the evidences of the Bible are so compelling, that copy errors and global thinking on behalf of the writers should have no adverse impact on one's faith, but they will keep the unbelieving skeptic from coming to the Lord by faith.

Chapter 26
The Bible and an evil world

By far the number one criticism leveled at the Bible and Christians who believe in it is if there is a loving God, then why is there so much evil in the world? With war, cancer, crime, disease, and oppression, the world is truly a mess. Yet, as painful as this world is, this question is far easier to answer than the question of why God chooses some people for salvation and not others (Rom. 9, Rom. 2:12-15), which is according to His own unrevealed counsel (Eph. 1:4-5).

God had warned man, that if he disobeyed by eating of the forbidden tree, he would die (Gen. 2:15-16). Therefore, death, destruction, disease, disasters, and dastardly deeds come through the fall of man. God was not joking when he warned him of the dire consequences that came from disobedience.

However, the odd accusations against God deal with crime and war. Crime is not a function of God harming and killing man, it is a function of man harming and killing man (Jam. 1:13-16). God has given man freewill. If man uses that freewill to hurt others, then man is to be blamed, not God.

You may argue that man should not have this kind of freewill. If so, man could not have any kind of freewill at all, since all disobedience is rebellion against God. Man could not cheat, lie, be unfaithful, or steal (a condition that will exist in the Millennium - Rev. 20-21, Is. 11, Ps. 72). Man would be forced to obey. If man is forced to obey, then he does not freely serve God. God allows man to choose his own course.

You might then respond that this is a high price to pay for freewill, and you would be correct, but this is what a fallen world apart from God looks like. In fact, if God had not put his restrainer, the Holy Spirit, on earth and in

96

believers, the violence of man would be so great that he would in short order destroy himself (Mt. 24, 2 Ths. 2, Gen. 11). God has limited man's actions, but still allows much freedom of action. So, it is not God, who is to blame for the cruelty in the world, but man.

If one allows questions about the cruelty of the world and the choices of God to stand in the way of faith, then one must deny the powerful evidence that is found in the Bible. To get into what is essentially God's business and not our own (Dt. 29:29) is to reject faith in favor of ignorance. Believers will understand all these issues one day as they stand before God (1 Cor. 13:12), but as for the time being, we take it by faith that all things will make sense at the proper time. Everyone who has rejected God because of faulty assumptions about the way He operates will certainly regret doing so, but by then it will be too late.

97

Chapter 27
It's all about the evidence

In the end everything is decided by the evidence. No matter how many respected experts believe something, tout something, or claim something, facts remain facts. The Bible tells us that if every person on earth believes something that is not true, then they all are liars and only God is truthful (Rom. 3:4). In this short treatise, we have viewed two basic kinds of evidence. First, we have adequately demonstrated that neither natural origins nor evolution has taken place or could ever have taken place. Second, we have demonstrated that the Bible is amply supported with an abundance of evidence, and is, therefore, the dependable source on which to base one's beliefs and life. Let's just summarize some of the fulfilled prophecies dealing with the coming of Jesus Christ as Jesus did for the disciples on the road to Emmaus (Lk. 24), so that we can reinforce your confidence in the Word:

OT Reference	Date	Prophecy	NT fulfillment
Gen. 3:15	1450 yrs. before Christ	A son would be born who would crush Satan's head	Mt. 26-28 Jesus' resurrection from the dead sealed Satan's fate

98

OT Reference	Date	Prophecy	NT fulfillment
Isaiah 7:14	700 years before Christ	Jesus would be born of a virgin and be called Immanuel (God with us)	Mt. 1, Lk. 2 Jesus is born of the Virgin Mary
Isaiah 9:6-7	700 years before Christ	Jesus would be the Everlasting Father (God Himself)	Jn. 20:28-31 Jesus receives worship from Thomas as the Lord and God
Micah 5:2	500 years before Christ	Jesus would be born in Bethlehem	Luke 2 Jesus is born in Bethlehem
Isaiah 40:3	700 years before Christ	The Voice Crying in the Wilderness prepared the way for the coming of Yahweh (the Lord God)	John 1, Mark 1 John the Baptist identifies himself as the Voice Crying in the Wilderness

			and Jesus as Yahweh
Psalm 22 (see Is. 53)	1000 years before Christ (and 700 years before the invention of crucifixion by the Carthigians)	This passage describes in detail the crucifixion, including such details as the soldiers casting lots for His clothes and that His bones would not be broken	Mt. 27, Luke 23, John 19 Jesus is crucified in the detail Ps. 22 states, including exact quotes (crucifixion was not invented until 250 BC)
Psalms 16:10	1000 years before Christ	Jesus' body would not see corruption (decay)	Mt. 27, Jn. 20, Lk. 24, Acts 1 Jesus was resurrected from the dead and now sits at the right hand of the Father

100

It is, therefore, incumbent upon all of us to trust the evidence and the common sense God has given to us, so that we can weed through all the hoopla bombarding us from unbelieving deceptive academics. I trust that this little volume has not only helped you fortify your faith, but has also provided some useful tools in your witnessing efforts. I would also encourage you again to view the DVDs *Unlocking the Mysteries of Life, Expelled: No Intelligence Allowed,* and *The Privileged Planet.* Just remember that these scientists are not Christians and are still trying to hold onto the very forms of naturalism that their research disproves. Read *Darwin's Black Box* and my two novels dealing in more detail with the same subjects, *The Group* and *The Debate.* Finally, trust your Bible and allow authors like Josh McDowell (*Evidence that Demands a Verdict, More Evidence that Demands a Verdict; More than a Carpenter*), Ravi Zacharius, Lee Strobel, C.S. Lewis, and Ken Ham to deepen your understanding so that you can become both informed and fully armed as a soldier and witness for Christ.

CONCLUSION

For the Christian college student, there is ample evidence to support his or her faith. This evidence is not only found in the Bible, but also in nature around us. Even the word *nature* means something that has been birthed (nativity is a related word). God has birthed the universe, life, and you.

For the unbeliever, there is plenty of evidence to find faith if you would only consider it. I trust that this little volume will encourage you in your search, fortify you when you find the truth, and enable you to defend your Christian faith.

If this little volume has whet your appetite for more evidence and a more detailed discussion of both the pros and cons of the Christian faith, I would encourage you to read my first two novels in the Danny Carter Series (The EU Trilogy), *The Group* and *The Debate*.

Blessings and Spiritual Vitality to you,
Dr. James L. Lowther
2 Timothy 3:15-16

THE GOSPEL OF JESUS CHRIST

1. God loves every person (Jn. 3:16 For God so loved the world...)

2. All men are sinners, falling short of God's perfection (Rm. 3)

3. God cannot allow sin or sinners into heaven (Ps. 5:4-5)

4. Souls of sinners are sent to an eternal Lake of Fire (Rv. 20)

5. God sent His Son, Jesus Christ to become a man and then die for the sins of people (Is. 7:14, 9:6,7, Jn. 3:16, Rm. 5)

6. Those who put their faith & trust in Jesus Christ are given eternal life (Rom. 10:9-13, Eph. 2:8-10, 1 Jn. 5:12-13)

For God So Loved the World that He gave His only begotten Son, that whosoever believes in Him shall not perish, but have eternal life. John 3:16

OTHER BOOKS BY DR. JAMES L. LOWTHER

THE DANNY CARTER SERIES (Novels)

The EU Trilogy: *The Group, The Debate; The Verdict*

Crossroad to Eternity

The Outcast

The Origins Science Textbook

Holding onto your Christian Faith while at the University

The Divine Webmaster

The God of Creation Daily Devotional

www.ingramcontent.com/pod-product-compliance
Lightning Source LLC
Chambersburg PA
CBHW051730170526
45167CB00002B/875